card & pocket on next page

# Rings with chain conditions

QA 251.4 .C47

Chatters, A. W.
Rings with chain conditions

(585686)

QA 251.4 .C47

Chatters, A. W.
Rings with chain conditions

(585686)

LIBRARY

MANKATO STATE UNIVERSITY

MANKATO, MINNESOTA

A W Chatters & C R Hajarnavis
University of Bristol/University of Warwick

# Rings with chain conditions



Pitman Advanced Publishing Program
BOSTON · LONDON · MELBOURNE

PITMAN PUBLISHING LIMITED
39 Parker Street, London WC2B 5PB

PITMAN PUBLISHING INC.
1020 Plain Street, Marshfield, Massachusetts

*Associated Companies*
Pitman Publishing Pty Ltd., Melbourne
Pitman Publishing New Zealand Ltd., Wellington
Copp Clark Pitman, Toronto



ISBN 0 273 08446 1

# Preface

This is an account of some of the recent developments in the theory of non-commutative rings with chain conditions. Often, but not always, we will be concerned with Noetherian rings, and we shall consider such aspects of their structure as decomposition theorems, the existence of quotient rings, and some special cases of the Jacobson conjecture. There are two themes which appear repeatedly throughout these notes. One is the use in proofs of Goldie's rank function for modules, and the other is the study of what can be thought of as the "Artinian part" of a Noetherian ring. There is a particular emphasis on new proofs and illustrative examples. The main classes of rings which are covered by the theory in these notes are the enveloping algebras of certain Lie algebras, Noetherian group rings such as the integral group ring of a polycyclic group, and various kinds of matrix rings.

The background material required is roughly what might be covered by an under-graduate or post-graduate course on non-commutative rings and their modules, including the Jacobson radical and the classical theory of semi-simple Artinian rings.

We hope that the reader will find the treatment fairly systematic.

A.W. Chatters

C.R. Hajarnavis

# Acknowledgements

We wish to express our sincere thanks and appreciation to Terri Moss for her excellent work in preparing the typescript; to Kenny Brown, Steve Ginn, Tom Lenagan and Toby Stafford for their encouragement and stimulating comments on the contents of these notes; and to David Griffel for all his invaluable help. Our thanks are also due to Andy Gray and Trevor Larkin for their assistance with the manuscript.

# Contents

# Conventions

"Ring" means associative but not necessarily commutative ring with identity element. A subring is not required to have an identity element. We shall usually ignore rings in which $1 = 0$.

References are given only in the remarks at the end of each chapter.

Numbers in round brackets such as (3.4) refer to the item of that number in these notes.

Unless otherwise stated Z and Q denote the ring of integers and the field of rational numbers respectively.

Unless otherwise stated N, or N(R), denotes the nilpotent radical of a ring R.

The abbreviations "a.c.c." and "d.c.c." stand for "ascending chain condition" and "descending chain condition" respectively.

A ring is said to be right Noetherian (or right Artinian) if it satisfies the a.c.c. (or d.c.c.) for right ideals.

$M^n$ denotes the direct sum of n copies of M.

The following will be standard notation for matrix rings:

$M_n(R)$ is the ring of all n by n matrices with entries in R;

$e_{ij}$ is the matrix with 1 in the (i,j)-position and 0's elsewhere (the size of $e_{ij}$ will be clear from the context);

$$\begin{pmatrix} A & B \\ C & D \end{pmatrix} = \{\begin{pmatrix} a & b \\ c & d \end{pmatrix} : a \in A, b \in B, c \in C, d \in D\}$$

and in this context $Be_{12}$ will denote the set of all 2 by 2 matrices with an arbitrary element of B in the (1,2)-position and 0's elsewhere, etc.

# 1 Goldie's theorem

The central aim of this chapter is to prove Goldie's theorem, but at the same time we shall introduce much of the terminology, notation, and basic material which will be used throughout these notes. The reader who is familiar with this subject may nevertheless find something of interest in this chapter; one of the proofs we give of Goldie's theorem is new, and there is a previously unpublished example due to G.M. Bergman of a primitive ring of Goldie dimension 1 which is not a Goldie ring (because of its length this example is put towards the end of the chapter in 1.36). We also give an example of J.W. Kerr which shows that a matrix ring over a commutative Goldie ring need not be a Goldie ring. As will be our usual practice, the chapter ends with some remarks which include references and other additional material.

Goldie's theorem gives very useful information about semi-prime rings which satisfy certain chain conditions and this, in turn, gives information about a ring R which has a nil (preferably nilpotent) semi-prime ideal N such that R/N satisfies the conditions for Goldie's theorem. In later sections of the chapter we introduce the concept of a quotient ring and interpret Goldie's theorem in that context, and we also prove some results about nil subrings being nilpotent.

We begin by introducing the concepts of right annihilator, essential right ideal, and right singular ideal. Let S be a non-empty subset of a ring R then the *right annihilator* of S in R is

$$r(S) = \{x \in R: sx = 0 \text{ for all } s \in S\};$$

the *left annihilator* of S is

$$\ell(S) = \{x \in R: xs = 0 \text{ for all } s \in S\}.$$

If S consists of a single element s we write $r(s)$ rather than $r(\{s\})$, etc. A right annihilator is a right ideal of R which is of the form $r(S)$ for some S, and a left annihilator is a left ideal of the form $\ell(S)$. The following basic properties of right and left annihilators are easy to prove: $S \subseteq T$ implies $r(T) \subseteq r(S)$ and $\ell(T) \subseteq \ell(S)$; $S \subseteq \ell(r(S)) \cap r(\ell(S))$; $r(\ell(r(S))) = r(S)$; $\ell(r(\ell(S))) = \ell(S)$. From these relationships it follows easily that the a.c.c. for right annihilators is equivalent to the d.c.c. for left annihilators. It is also easy to show that if R satisfies the a.c.c. or d.c.c. for right annihilators then so also does any subring of R.

Let M be a right R-module. A submodule K of M is said to be *essential* in M or to be an essential submodule of M if $K \cap A \neq 0$ whenever A is a non-zero submodule of M. In these circumstances we also say that M is an essential extension of K. It is easy to show that the intersection of a finite number of essential submodules of M is essential in M, and any submodule of M which contains an essential submodule of M is itself essential in M. Also if K is an essential submodule of L and L is an essential submodule of M then K is essential in M. A right ideal of R is said to be essential if it is essential as a right R-submodule of R, i.e. if it has non-zero intersection with each non-zero right ideal of R. For example, in the ring Z of integers every non-zero ideal is essential because each pair of non-zero ideals of Z has non-zero intersection. On the other hand, if R is a semi-simple Artinian ring and I is a right ideal of R then I = eR for some idempotent element e of R and we have $I \cap (1 - e)R = 0$, so that I is essential only if $(1 - e)R = 0$, i.e. only if I = R.

Lemma 1.1. Let M be a right R-module, let a be a non-zero element of M and let K be an essential submodule of M, then there is an essential right ideal L of R such that $aL \neq 0$ and $aL \subseteq K$.

Proof. Set $L = \{r \in R: ar \in K\}$, then L is a right ideal of R and $aL \subseteq K$. We have $aR \cap K \neq 0$ so that ar is a non-zero element of K for some $r \in R$; in fact $r \in L$ so that $aL \neq 0$. Now let I be a non-zero right ideal of R. We wish to show that $I \cap L \neq 0$ and this is certainly the case if $aI = 0$ because then $I \subseteq L$. Suppose now that $aI \neq 0$, then $aI \cap K \neq 0$. Thus ax is a non-zero element of K for some $x \in I$. We have $x \in L$, so that $I \cap L \neq 0$. □

The following is a useful way of constructing essential submodules of a module M. Let A and B be submodules of M with $A \cap B = 0$. Zorn's lemma can be applied to the set of submodules of M which contain B and have zero intersection with A to obtain a submodule C of M such that $A \cap C = 0$, $B \subseteq C$, and $A \oplus C$ is an essential submodule of M. The next result is an illustration of how this method can be used. We define soc(M) the *socle* of a right R-module M, to be the sum of all the simple submodules of M (with soc(M) = 0 if M has no simple submodules). A Zorn's lemma argument shows that soc(M) is a direct sum of simple submodules.

Lemma 1.2. Let E be the intersection of all the essential submodules of a module M, then soc(M) = E.

Proof. Concerning the definition of E we note that M has at least one essential submodule, namely M itself. If S and K are submodules of M such that S is simple and K is essential in M then $S \cap K \neq 0$, so that $S \cap K$ is a non-zero submodule of the simple module S. It follows that $S \subseteq K$ and that $\mathrm{soc}(M) \subseteq E$. To derive the reverse inclusion we shall show that E is a sum of simple modules by proving the equivalent condition that each submodule of

E is a direct summand of E. (The case E = 0 is trivial).

Let A be any submodule of E. There is a submodule C of M such that $A \cap C = 0$ and $A \oplus C$ is an essential submodule of M. We have $E \subseteq A \oplus C$. Set $D = E \cap C$. By the modular law we have $E = E \cap (A \oplus C) = A \oplus (E \cap C) = A \oplus D$.□

The *right singular ideal* Z(R) of a ring R is defined by

$$Z(R) = \{r \in R: rK = 0 \text{ for some essential right ideal } K \text{ of } R\}.$$

In other words, if $x \in R$ then $x \in Z(R)$ if and only if $r(x)$ is an essential right ideal of R. It follows from the material already given that Z(R) is a two-sided ideal of R. For example, let $z \in Z(R)$ and $a \in R$. Because $r(z)$ is essential there is an essential right ideal L of R such that $aL \subseteq r(z)$ (1.1); (if a = 0 we can take L = R). Thus zaL = 0 so that $za \in Z(R)$. The *singular submodule* of a module is defined similarly.

<u>Lemma 1.3</u>. Let R be a commutative ring, then the singular ideal Z(R) of R is zero if and only if R is semi-prime.

<u>Proof</u>. Suppose that R is semi-prime and let $z \in Z(R)$. Set $I = zR \cap r(z)$. We have $zR.r(z) = 0$. Also $I^2 \subseteq zR.r(z)$ from which it follows that I = 0. But $r(z)$ is an essential ideal of R. Hence zR = 0, i.e. z = 0.

Conversely suppose that Z(R) = 0 and let a be an element of R such that $a^2 = 0$. We shall show that a = 0, from which it follows that R has no non-zero nilpotent elements. Let x be a non-zero element of R, then either ax = 0 in which case $x \in r(a)$, or $ax \neq 0$ in which case ax is a non-zero element of $r(a)$. Thus $xR \cap r(a) \neq 0$. Therefore $r(a)$ is an essential ideal of R, so that a = 0. □

Example 1.4. Let R be the ring of 2 by 2 upper triangular matrices over a field F. The right socle $Fe_{12} + Fe_{22}$ of R has zero left annihilator, from which it follows that the right singular ideal of R is zero. However, R is not semi-prime because the strictly upper triangular elements of R form a non-zero nilpotent ideal. □

A ring R which has Z(R) = 0 is called *right non-singular*. Thus a commutative ring is non-singular if and only if it is semi-prime (1.3). A right non-singular ring need not be semi-prime (1.4), and a prime ring need not be right non-singular (1.36), but many semi-prime rings are right non-singular (1.6). The *left singular ideal* of R is the set of all elements of R whose left annihilators are essential left ideals, and in general the right and left singular ideals do not coincide.

Example 1.5. Let F = Z/2Z and set

$$R = \begin{bmatrix} Z & F \\ 0 & F \end{bmatrix} .$$

The right ideals $Fe_{12}$ and $Fe_{22}$ are simple right R-modules and hence are contained in each essential right ideal of R (1.2). Also $2nZe_{11}$ is a right ideal of R for each integer n. It follows easily that a right ideal of R is essential if and only if it contains $2nZe_{11} + Fe_{12} + Fe_{22}$ for some non-zero integer n. Each of these right ideals has zero left annihilator, so that the right singular ideal of R is zero. Set $L = 2Ze_{11} + Fe_{12} + Fe_{22}$. Let r be a non-zero element of R, then either the (1,1)-entry of r is zero in which case $r \in L$, or it is non-zero in which case 2r is a non-zero element of L. Thus in either case we have $Rr \cap L \neq 0$, so that L is an essential left ideal of R. But $Le_{12} = 0$ so that the left singular ideal of R is non-zero. In fact the left singular ideal of R is $Fe_{12}$, which is nilpotent. □

The next few results are of independent interest besides being part of the preparations for Goldie's theorem.

Theorem 1.6. (Mewborn and Winton) Let R be a ring with the a.c.c. for right annihilators, then the right singular ideal of R is nilpotent.

Proof. In this proof we shall write Z rather than Z(R) for the right singular ideal of R. Because $Z \supseteq Z^2 \supseteq Z^3 \supseteq \ldots$ we have $r(Z) \subseteq r(Z^2) \subseteq r(Z^3) \subseteq \ldots$. Therefore there is a positive integer n such that $r(Z^n) = r(Z^{n+1})$. Suppose that $Z^{n+1} \neq 0$; we shall obtain a contradiction. There is an element a of Z such that $Z^n a \neq 0$. Choose such an element a with $r(a)$ as large as possible. Let $b \in Z$ then $r(b)$ is an essential right ideal, so that $r(b) \cap aR \neq 0$. Thus there is an element r of R such that $ar \neq 0$ and $ar \in r(b)$. We have $ba \in Z$ and $r(a) \subseteq r(ba)$. But $ar \neq 0$ and $bar = 0$. Therefore $r(a)$ is strictly contained in $r(ba)$. It follows from the choice of a that $Z^n ba = 0$. But b is an arbitrary element of Z. Hence $Z^{n+1}a = 0$, so that $Z^n a = 0$. □

Theorem 1.7. Let R be a semi-prime ring with the a.c.c. for right annihilators, then R has no non-zero nil one-sided ideals.

Proof. Let I be a non-zero one-sided ideal of R and let a be a non-zero element of I with $r(a)$ as large as possible. Because R is semi-prime there is an element x of R such that $axa \neq 0$. Thus axa is a non-zero element of I with $r(a) \subseteq r(axa)$, so that $r(a) = r(axa)$. We have $ax \neq 0$, i.e. $x \notin r(a)$. Thus $x \notin r(axa)$, so that $(ax)^2 \neq 0$. Hence $xax \notin r(a)$ so that $(ax)^3 \neq 0$. And so on. Therefore ax, and hence also xa, is not nilpotent and ax or xa belongs to I. □

Let A,B be nilpotent right ideals of a ring R. Then $A^k = B^n = 0$ for some integers k and n. It is easy to see that $(A + B)^{k+n} = 0$ since $(A + B)^{k+n}$

is a sum of products in which either A occurs at least k times or B occurs at least n times. It follows that the sum of a finite number of nilpotent right ideals is nilpotent. Notice also that if A is a nilpotent right ideal then RA is also nilpotent. Thus every nilpotent one-sided ideal is contained in a nilpotent (two-sided) ideal.

Corollary 1.8. (Levitzki) Let R be a right Noetherian ring then each nil one-sided ideal of R is nilpotent.

Proof. Let W be the sum of all the nilpotent right ideals of R. Then W is an ideal and because R is right Noetherian, W is the sum of a finite number of nilpotent right ideals and hence is nilpotent. It follows easily that R/W has no non-zero nilpotent right ideals. Let I be a nil one-sided ideal of R, then the image of I in R/W is zero (1.7), so that $I \subseteq W$. □

We next give some more definitions which will be needed for Goldie's theorem. An element c of a ring R is *right regular* if $r(c) = 0$, *left regular* if $\ell(c) = 0$, and *regular* if $\ell(c) = r(c) = 0$. For example, every non-zero element of an integral domain is regular, and if F is a field then an element of $M_n(F)$ is regular if and only if its determinant is non-zero. Elements which are regular on one side need not be regular. In 1.5 the element $2e_{11}+e_{22}$ is left regular but not right regular. If I is an ideal of R we set

$$C(I) = \{c \in R: c + I \text{ is a regular element of } R/I\}$$

and we write C(0) for the set of regular elements of R.

We next introduce the concept of Goldie dimension (also known as uniform dimension). A non-zero module U is said to be *uniform* if any two non-zero submodules of U have non-zero intersection, i.e. if each non-zero submodule of U is essential in U. Let M be an R-module. We say that M has *finite Goldie dimension* if M does not contain a direct sum of an infinite number of non-zero submodules. It is easy to show that M has finite Goldie dimension if M is Noetherian or Artinian. A ring R is said to have *finite right Goldie dimension* if R has finite Goldie dimension as a right R-module. We call R a *right Goldie ring* if it has finite right Goldie dimension and satisfies the a.c.c. for right annihilators. A right Noetherian ring is right Goldie, but the converse is not true because any commutative integral domain is trivially a Goldie ring.

The next lemma is not needed to prove Goldie's theorem but it gives the basic properties of modules of finite Goldie dimension.

Lemma 1.9. Let M be a non-zero right R-module.

(a) If M has finite Goldie dimension then each non-zero submodule of M contains a uniform submodule, and there is a finite number of uniform submodules of M whose sum is direct and is an essential submodule of M.

(b) Suppose that M has uniform submodules $U_1,\ldots,U_n$ such that the sum $U_1 + \ldots + U_n$ is direct and is an essential submodule of M, then M has finite Goldie dimension and the positive integer n is independent of the choice of the $U_i$. We call n the *Goldie dimension* of M and denote it by dim M.

Proof. (a) Suppose that M has finite Goldie dimension and let K be a non-zero submodule of M. We wish to show that K has a uniform submodule. This is trivial if K is uniform. Suppose that K is not uniform, then there are non-zero submodules $A_1$ and $B_1$ of K such that $A_1 \cap B_1 = 0$. Thus $A_1 + B_1$ is the direct sum of two non-zero submodules of M. If $A_1$ or $B_1$ is uniform we can stop. Otherwise there are non-zero submodules $A_2$ and $B_2$ of $B_1$ such that $A_2 \cap B_2 = 0$. Thus $A_1 + A_2 + B_2$ is the direct sum of three non-zero submodules of M. Because M has finite Goldie dimension this process must stop after a finite number of steps, and when it does it gives a uniform submodule of K. In particular, M has a uniform submodule $U_1$. Suppose that $U_1$ is not essential in M then there is a non-zero submodule $K_1$ of M such that $U_1 \cap K_1 = 0$. Let $U_2$ be a uniform submodule of $K_1$, then the sum $U_1+U_2$ is direct. If $U_1 \oplus U_2$ is not essential in M then there is a non-zero submodule $K_2$ of M such that $(U_1 \oplus U_2) \cap K_2 = 0$, and so on. This process also must stop after a finite number of steps.

(b) Suppose that $U_1,\ldots,U_n$ are uniform submodules of M such that the sum $U_1 + \ldots + U_n$ is direct and is essential in M. The first step is to suppose that K is a submodule of M such that, for each i, we have $K \cap U_i \neq 0$ and to prove that K is essential in M. For each i set $V_i = K \cap U_i$ and let $L = V_1 + \ldots + V_n$. We have $L \subseteq K$ so that it is enough to show that L is essential in M. Let x be a non-zero element of M. We have $xR \cap (U_1 \oplus \ldots \oplus U_n) \neq 0$. Thus there is an element r of R such that $xr = u_1 + \ldots + u_n$ where $u_i \in U_i$ and say $u_1 \neq 0$.

We have $V_1$ essential in $U_1$ because $V_1 \neq 0$. Now there is an essential right ideal $X_1$ of R such that $u_1X_1$ is a non-zero submodule of $V_1$ (1.1). Thus there is an element $r_1$ of R such that $xr_1 = v_1 + a_2 + \ldots + a_n$ for some $v_1 \in V_1$ and $a_i \in U_i$ with $v_1 \neq 0$. If $a_2 = \ldots = a_n = 0$ we can stop because then $xr_1$ is a non-zero element of $xR \cap L$. If say $a_2 \neq 0$ then there is an essential right ideal $X_2$ of R such that $a_2X_2$ is a non-zero submodule of $V_2$ (1.1); we may have $v_1X_2 = 0$ but that does not matter. Thus there is an element $r_2$ of R such that $xr_2 = w_1 + w_2 + b_3 + \ldots + b_n$ for some $w_i \in V_i$ and $b_i \in U_i$ with $w_2 \neq 0$. Eventually we obtain a non-zero element of $xR \cap L$.

Now suppose that $Y_1,\ldots,Y_k$ are non-zero submodules of M such that the sum $Y_1 + \ldots + Y_k$ is direct. We shall show that $k \leq n$ from which (b) follows easily. Set $W = Y_2 + \ldots + Y_k$ then W is not essential in M because $W \cap Y_1 = 0$. Therefore $W \cap U_i = 0$ for some i and we may suppose that $W \cap U_1 = 0$. Thus the sum $U_1 + Y_2 + \ldots + Y_k$ is direct. Similarly $U_1 + Y_3 + \ldots + Y_k$ is not essential so that we have say $(U_1 + Y_3 + \ldots + Y_k) \cap U_2 = 0$. Thus the sum

$U_1 + U_2 + Y_3 + \ldots + Y_k$ is direct. In this way we obtain $k \leq n$. □

Let M be a module of finite Goldie dimension. It is clear from the definition that submodules of M also have finite Goldie dimension, but it is not always true that arbitrary factor modules of M have finite Goldie dimension. For example, Q has Goldie dimension 1 as a Z-module because any two non-zero additive subgroups of Q have non-zero intersection, but Q/Z does not have finite Goldie dimension. If V is a vector space then V has finite Goldie dimension if and only if V has finite dimension in the usual sense of linear algebra, and in these circumstances the two dimensions are equal.

Let $M_1$ and $M_2$ be right R-modules of finite Goldie dimension n and k with essential submodules $E_1$ and $E_2$ respectively. Let x be a non-zero element of $M_1 \oplus M_2$. Then using (1.1) it is not difficult to show that $xR \cap (E_1 \oplus M_2) \neq 0$. Since $E_1 \oplus E_2 = (M_1 \oplus E_2) \cap (M_2 \oplus E_1)$ it follows that $E_1 \oplus E_2$ is essential in $M_1 \oplus M_2$. Thus $M_1 \oplus M_2$ has dimension n+k (1.9(b)).

We are now ready to prove Goldie's theorem. We shall give a quick proof (also due to Goldie) in 1.10, and in 1.19 we shall give another proof which is harder but which achieves more.

Theorem 1.10. (Goldie) Let R be a semi-prime right Goldie ring and let I be an essential right ideal of R, then I contains a regular element of R.

Proof. We shall show that R contains a right regular element and then apply 1.13. We know that I is not nil (1.7). Let $a_1$ be a non-nilpotent element of I with $r(a_1)$ as large as possible. We have $r(a_1) \subseteq r(a_1^2)$ and $a_1^2$ is a non-nilpotent element of I. By choice of $a_1$ we have $r(a_1) = r(a_1^2)$. If $r(a_1) = 0$ we stop. If not, we have $r(a_1) \cap I \neq 0$. Let $a_2$ be a non-nilpotent element of $r(a_1) \cap I$ with $r(a_2)$ as large as possible. Then $r(a_2) = r(a_2^2)$. Let

$a_1r_1 = a_2r_2$ for some $r_1, r_2 \in R$. Because $a_1a_2 = 0$ we have $a_1^2r_1 = 0$. Therefore $a_1r_1 = 0$. Hence the sum $a_1R + a_2R$ is direct. The same argument shows that $r(a_1 + a_2) = r(a_1) \cap r(a_2)$. If $r(a_1 + a_2) = 0$ we stop. If not, let $a_3$ be a non-nilpotent element of $r(a_1 + a_2) \cap I$ with $r(a_3)$ as large as possible. Then $r(a_3) = r(a_3^2)$ and the sum $a_1R + a_2R + a_3R$ is direct (because $a_1a_2 = a_1a_3 = a_2a_3 = 0$). Thus $r(a_1 + a_2 + a_3) = r(a_1) \cap r(a_2) \cap r(a_3)$. Because R has finite right Goldie dimension this process must stop after a finite number of steps, and when it does there are elements $a_1, \ldots, a_n$ of I such that $r(a_1 + \ldots + a_n) = 0$ and we use 1.13. □

Lemma 1.11. Let R be a ring with finite right Goldie dimension and let c be a right regular element of R, then cR is an essential right ideal of R.

Proof. Let I be a right ideal of R with $I \cap cR = 0$ then the sum $I + cI + c^2I + c^3I + \ldots$ is direct. Therefore $c^nI = 0$ for some n, and because c is right regular it follows that $I = 0$. □

Lemma 1.12. Let R be a right non-singular ring with finite right Goldie dimension then right regular elements of R are regular.

Proof. Let c be a right regular element of R then cR is an essential right ideal (1.11). But $\ell(c) = \ell(cR)$ and R is right non-singular. Therefore $\ell(c) = 0$. □

Corollary 1.13. Let R be a semi-prime right Goldie ring then right regular elements of R are regular.

Proof. Combine 1.6 and 1.12. □

Lemma 1.14. Let R be a right non-singular ring with finite right Goldie dimension then R satisfies the a.c.c. and d.c.c. for right annihilators.

*Proof*. Let A and B be right annihilators in R with $A \subseteq B$. Suppose that A is an essential submodule of B. We shall show that A = B. Let $b \in B$ then there is an essential right ideal L of R such that $bL \subseteq A$ (1.1). Thus $\ell(A)bL = 0$. But R is right non-singular. Hence $\ell(A)b = 0$ so that $b \in r(\ell(A))$, i.e. $b \in A$.

Thus if A and B are right annihilators in R and A is strictly contained in B then there is a non-zero right ideal C of R such that $C \subseteq B$ and $A \cap C = 0$. It follows that a chain of distinct right annihilators in R gives rise to a direct sum of non-zero right ideals such as C. Therefore R has the a.c.c. and d.c.c. for right annihilators. □

Corollary 1.15. A semi-prime right Goldie ring has the d.c.c. for right annihilators.

*Proof*. Combine 1.6 and 1.14. □

Let R be a semi-prime ring and let A and B be right ideals of R with AB = 0, then $(BA)^2 = 0$ and $(A \cap B)^2 = 0$ so that BA = 0 and $A \cap B = 0$. Thus if I is an ideal of R then $Ir(I) = 0$ so that $r(I)I = 0$. Similarly $I\ell(I) = 0$. Therefore $\ell(I) = r(I)$. If I is a right annihilator then $I = r(\ell(I)) = \ell(r(I))$ so that I is also a left annihilator, and in these circumstances we call I an *annihilator ideal*.

A prime ideal P of a ring R is called a *minimal prime* if P does not properly contain any prime ideal of R. To put the next result into context we note that if R is any ring with a prime ideal P then Zorn's lemma can be applied to the set of primes contained in P to show that P contains a minimal prime. In what follows we will not consider R to be a prime ideal of R.

Lemma 1.16. Let R be a semi-prime ring with the a.c.c. (equivalently d.c.c.) for annihilator ideals, then R has only a finite number of minimal prime ideals. If $P_1,\ldots,P_n$ are the minimal prime ideals of R then $P_1 \cap \ldots \cap P_n = 0$. Also a prime ideal of R is minimal if and only if it is an annihilator ideal.

Proof. By an "annihilator prime" we mean a prime ideal which is an annihilator. We shall first show that every annihilator ideal of R contains a product of annihilator primes. Suppose not, then there is an annihilator ideal A which is maximal with respect to not containing a product of annihilator primes. Because A cannot itself be prime, there are ideals B and C of R which strictly contain A such that $BC \subseteq A$. We have $\ell(A)BC = 0$ so that we may take $C = r(\ell(A)B)$. Similarly we can take $B = \ell(Cr(A))$. Thus B and C are annihilator ideals which strictly contain A and $BC \subseteq A$. Therefore B and C each contain a product of annihilator primes and hence so also does A. This is a contradiction.

Because the zero ideal of R is an annihilator, there are annihilator primes $P_1,\ldots,P_n$ of R such that $P_1P_2\ldots P_n = 0$. We have $(P_1 \cap \ldots \cap P_n)^n \subseteq P_1P_2\ldots P_n$ from which it follows that $P_1 \cap \ldots \cap P_n = 0$. If P is a minimal prime ideal of R then $P_1P_2 \ldots P_n \subseteq P$ so that $P_i \subseteq P$ for some i. Hence $P = P_i$.

To complete the proof we assume that P is an annihilator prime ideal and show that P is a minimal prime ideal. Let P' be a prime ideal of R with $P' \subseteq P$. Suppose that $\ell(P) \subseteq P'$ then $\ell(P) \subseteq P$ so that $(\ell(P))^2 = 0$. Hence $\ell(P) = 0$. Therefore $P = R$, this is not allowed. Thus $\ell(P)P \subseteq P'$ and P' does not contain $\ell(P)$. Therefore $P \subseteq P'$ so that $P' = P$. □

The proof of 1.16 can easily be modified to show that every ideal I of a right Noetherian ring contains a product of prime ideals all of which contain I.

Let R be a right Noetherian ring with nilpotent radical N. Let $P_1,\ldots,P_k$ be the minimal prime ideals of R. It follows immediately from 1.16 that $N=P_1 \cap \ldots \cap P_k$. It is also not difficult to show that $C(N)=C(P_1)\cap\ldots\cap C(P_k)$.

The reader who does **not** wish at the moment to go through the following generalisation and second proof of Goldie's theorem should go to 1.22.

Lemma 1.17. Let R be a right non-singular ring with the a.c.c. for right annihilators and let c be an element of R such that cR is an essential right ideal, then c is right regular.

Proof. Let $x \in R$ and let A and B be right ideals of R such that A is an essential right R-submodule of B. We shall show that xA is essential in xB. Let $b \in B$ with $xb \neq 0$, then there is an essential right ideal L of R such that $bL \subseteq A$ (1.1). Because R is right non-singular we have $xbL \neq 0$. Also $xbL \subseteq xbR \cap xA$ from which it follows that xA is essential in xB.

Taking x = c with A = cR and B = R we see that $c^2R$ is an essential right ideal of R. Similarly $c^kR$ is essential for each positive integer k. But $r(c) \subseteq r(c^2) \subseteq r(c^3) \subseteq \ldots$ so that $r(c^n) = r(c^{n+1})$ for some n. This implies that $c^nR \cap r(c) = 0$. Thus $r(c) = 0$. □

The proof of the next lemma is an adaptation of an argument of A. Ludgate.

Lemma 1.18. Let R be a prime ring with the a.c.c. and d.c.c. for right annihilators, let I be an essential right ideal of R and let $a \in R$, then a + I contains a regular element of R where $a + I = \{a + x: x \in I\}$.

Proof. Let $x \in I$ with $r(a + x)$ as small as possible and set c = a + x. Let B be a right ideal of R with $B \cap cR = 0$. Let $b \in B \cap I$ then $c + b \in a+I$. Because $cR \cap bR = 0$ we have $r(c + b) = r(c) \cap r(b)$ so that $r(c + b) \subseteq r(c)$. By choice of c we have $r(c + b) = r(c)$. Hence $r(c) \subseteq r(b)$ for all $b \in B \cap I$.

Therefore $(B \cap I)r(c) = 0$. Because R is a prime ring we have $r(c) = 0$ or $B \cap I = 0$. If $B \cap I = 0$, then $B = 0$ because I is essential. It follows that cR is essential. Therefore $r(c) = 0$ (1.17 and 1.6).

At this stage we know that $a + I$ contains a right regular element. Let d be a right regular element of $a + I$ with $\ell(d)$ as small as possible. There is a right R-submodule Y of I such that $Y \cap dR = 0$ and $dR \oplus Y$ is essential in R. Let A be a left ideal of R with $A \cap Rd = 0$ and let $y \in A \cap Y$. We have $Rd \cap Ry = 0$ so that $\ell(d + y) = \ell(d) \cap \ell(y)$. Also $d + y \in a + I$, and because $dR \cap yR = 0$ we have $r(d + y) = r(d) \cap r(y) = 0$. Therefore by choice of d we have $\ell(d) = \ell(d + y)$ so that $\ell(d) \subseteq \ell(y)$. Hence $\ell(d)(A \cap Y) = 0$ so that $\ell(d)YA = 0$. Therefore $A\ell(d)Y = 0$ so that $A\ell(d)(dR \oplus Y) = 0$. But $dR \oplus Y$ is an essential right ideal of R. Also R is right non-singular (1.6). Therefore $A\ell(d) = 0$ so that $A = 0$ or $\ell(d) = 0$. We can now use the argument at the end of the last paragraph to show that $\ell(d) = 0$. □

We remark that if R is a prime right Goldie ring then $r(c) = 0$ implies that cR is essential (1.11) and so $\ell(c) = 0$ as R is non-singular. Thus the argument of the previous paragraph is not required in this case.

Theorem 1.19. Let R be a semi-prime ring with the a.c.c. and d.c.c. for right annihilators, let I be an essential right ideal of R and let $a \in R$, then $a + I$ contains a regular element of R.

Proof. Let $P_1, \ldots, P_n$ be the distinct minimal prime ideals of R then $P_1 \cap \ldots \cap P_n = 0$ and each $P_i$ is an annihilator ideal (1.16). For each i set $A_i = \ell(P_i) = r(P_i)$, then $A_i$ is the intersection of those $P_j$ with $j \neq i$. Let X be a left ideal of R with $P_i \subseteq X$ for some i, and let Y be the right ideal of R such that $P_i \subseteq Y$ and $Y/P_i$ is the right annihilator in $R/P_i$ of $X/P_i$. It is straightforward to show that $Y = r(A_iX)$ and hence that $R/P_i$

satisfies the a.c.c. and d.c.c. for right annihilators for all i.

Set $B_i = I \cap A_i$. We shall show that $(B_i + P_i)/P_i$ is essential as a right ideal of $R/P_i$. Let K be a right ideal of R such that $K \cap B_i \subseteq P_i$ then $K \cap B_i = 0$ because $B_i \cap P_i = 0$. Thus $K \cap I \cap A_i = 0$ so that $K \cap A_i = 0$. Hence $K \subseteq \ell(A_i)$, i.e. $K \subseteq P_i$. It follows that $(B_i + P_i)/P_i$ is an essential right ideal of $R/P_i$. Therefore there exists $x_i \in B_i$ such that $a + x_i \in C(P_i)$ (by 1.18 applied in $R/P_i$). Note that $x_i \in I \cap P_j$ for all $j \neq i$. Set $c = a + x_1 + \dots + x_n$ then $c \in a + I$. Let $y \in R$ with $cy = 0$ then, in particular we have $cy \in P_1$. But $x_i y \in P_1$ for all $i \neq 1$. Therefore $(a + x_1)y \in P_1$. But $a + x_1 \in C(P_1)$ so that we have $y \in P_1$. Similarly $y \in P_i$ for all i so that $y = 0$. Thus $r(c) = 0$. Similarly $\ell(c) = 0$. □

The special case of 1.19 in which $a + I = I$ was proved by Johnson and Levy and independently by Hajarnavis and Ludgate.

<u>Corollary 1.20</u>. (Robson) Let R be a semi-prime right Goldie ring. Let I be an essential right ideal of R and let $a \in R$, then $a + I$ contains a regular element of R.

<u>Proof</u>. Combine 1.19 and 1.15. □

<u>Theorem 1.21</u>. (Robson) Let R be a semi-prime right Goldie ring and let I be an essential right ideal of R, then I is generated as a right ideal by the set of regular elements of R which belong to I.

<u>Proof</u>. Let T be the right ideal of R generated by the regular elements which belong to I. We know that I contains at least one regular element c (1.10). Because cR is an essential right ideal of R (1.11) so also is T. Let $a \in I$ then $a + t$ is regular for some $t \in T$ (1.19 and 1.15). Hence $a + t \in T$ so that $a \in T$. □

The following example shows that a one-sided regular element of a prime ring with the a.c.c. and d.c.c. for right annihilators need not be regular (cf. 1.13 and the proof of 1.18).

Example 1.22. Let R be a right Noetherian integral domain which has non-zero elements a and b such that $Ra \cap Rb = 0$ (we will give a specific example in a moment). The ring $M_2(R)$ is prime and right Noetherian and the element $ae_{11} + be_{21}$ is left regular but not regular.

To construct such a ring R, let F be a field and let x and y be indeterminates. Let K be the field of rational functions in y over F, and let s be the endomorphism of K determined by $s(y) = y^2$. Then s is not surjective. Let R be the set of polynomials in x with coefficients in K and with the coefficients written on the right of the powers of x. We make R into a ring by adding elements of R in the usual way, and by defining $kx = xs(k)$ for all $k \in K$ and extending this in the obvious way to give the product of any two elements of R. As far as right ideals of R are concerned, the theory is like that for commutative polynomials and every right ideal of R is principal. On the other hand, if we set $a = x$ and $b = xy$ then $Ra \cap Rb = 0$.

It is worth noting the sum $Rxy + Rxyx + Rxyx^2 + \dots + Rxyx^n + \dots$ is direct, and in fact any integral domain which does not have left Goldie dimension 1 can be shown in a similar way not to have finite left Goldie dimension. □

It is easy to find examples of semi-prime rings with the a.c.c. for right annihilators which are not right Goldie rings (e.g. the opposite ring of 1.22). It is much harder to find examples of semi-prime rings with finite right Goldie dimension which are not right Goldie rings, but such an example will be given in 1.36.

Before re-interpreting Goldie's theorem in terms of quotient rings we shall derive some very important consequences of the form given in 1.10.

The right socle soc(R) of a ring R is the socle of the right R-module R, i.e. soc(R) is the sum of all the minimal right ideals of R, (with soc(R) = 0 if R has no minimal right ideals). It is easy to show that soc(R) is an ideal of R, and we know that soc(R) coincides with the intersection of all the essential right ideals of R (1.2). In general, the right and left socles of R do not coincide. Let R be a semi-prime ring and let K be a minimal right ideal of R, then K = eR for some idempotent e. Also if f is any idempotent element of R then fR is a minimal right ideal if and only if fRf is a division ring, so that fR is a minimal right ideal if and only if Rf is a minimal left ideal (we leave the proofs as an exercise). It follows that the left and right socles of a semi-prime ring coincide.

Theorem 1.23. Let R be a semi-prime right Goldie ring and let S be the socle of R, then there is a central idempotent element e of R such that S = eR.

Proof. There is a right ideal T of R such that $S \cap T = 0$ and $S \oplus T$ is an essential right ideal (cf. paragraph after 1.1). There exist $s \in S$ and $t \in T$ such that s + t is regular (1.10). Set c = s + t. Because S is an ideal of R and $S \cap T = 0$ we have TS = 0. Left multiplication by c induces an injective endomorphism of the semi-simple right R-module S. We know that S has finite length as a right R-module because R has finite right Goldie dimension. Therefore S = cS. Thus s = ce for some $e \in S$. We have $ce = se + te = se = ce^2$ so that $e = e^2$. Also for any $x \in S$ we have cx = sx + tx = sx = cex so that x = ex. Thus S = eR. To show that e is central, let $y \in R$ then $ye \in S$ and S = eR so that ye = eye. Also ey(1 - e)Rey(1 - e) = 0 because $Re \subseteq eR$. Hence ey(1 - e) = 0 so that ey = eye. □

Theorem 1.24. Let R be a prime right Goldie ring and assume that the socle S of R is non-zero, then R is a simple Artinian ring.

Proof. Let A be a non-zero ideal of R and suppose that I is a right ideal of R with $I \cap A = 0$. Then $IA = 0$. Because R is prime it follows that $I = 0$. Thus every non-zero ideal of a prime ring is essential as a one-sided ideal. Hence $S \subseteq A$ (1.2). Also S contains a regular element c (1.10). As in the proof of 1.23 we have $S = cS$. But $cR \subseteq S$. Therefore $cR = cS$. Hence $R = S$. Thus $A = R$ so that R is simple, and $R = S$ so that R is Artinian. □

Theorem 1.25. Let R be a ring such that the nil radical N of R is nilpotent and R/N is a right Goldie ring, then a prime ideal of R is minimal if and only if it does not contain an element of C(N).

Proof. Without loss of generality we may suppose that R is a semi-prime right Goldie ring because all prime ideals of R contain N. We must prove that a prime ideal of R is minimal if and only if it does not contain a regular element. Let P be a minimal prime of R then P is an annihilator (1.16) so that P does not contain a regular element.

Now suppose that P is a prime ideal of R which is not minimal. We shall show that P is essential as a right ideal. There is a prime ideal P' strictly contained in P. Let I be a right ideal of R with $I \cap P = 0$. Then $IP = 0$ so that $IP \subseteq P'$. Hence $I \subseteq P'$ from which it follows that $I = 0$. Thus P is an essential right ideal of R so that P contains a regular element (1.10). □

## Quotient Rings

Let R be any ring. The *right quotient ring* of R, if it exists, is a ring Q which satisfies the following conditions:

(i) R is a subring of Q;

(ii) Each regular element of R is a unit of Q;

(iii) Each element q of Q is of the form $ac^{-1}$ for some elements a and c of R with c regular (i.e. $qc \in R$ for some regular $c \in R$).

In this situation we say that R is a *right order* in Q. The right quotient ring of R, if it exists, is unique in the sense that any two such rings are isomorphic under a homomorphism which restricts to the identity function on R. The left quotient ring of R is defined similarly (with $c^{-1}a$ instead of $ac^{-1}$). If both the left and right quotient rings exist then, as we shall show in a moment, they are equal and we speak of "*the quotient ring of* R". The kind of quotient ring which we are considering here is sometimes called the *classical quotient ring* to distinguish it from other kinds.

Let R be a ring which has a right quotient ring Q, then Q is a *quotient ring* in the sense that every regular element of Q is a unit of Q (i.e. Q is its own quotient ring). For let $q = ac^{-1}$ be a regular element of Q where a, $c \in R$ with c regular. To show that q is a unit of Q it is enough to show that a is a regular element of R. Let $x \in R$ with $xa = 0$ then $xq = 0$ so that $x = 0$. Now let $y \in R$ with $ay = 0$. We have $a = qc$ so that $qcy = 0$. Therefore $cy = 0$ so that $y = 0$.

We shall now show that a right regular element c of a right Artinian ring R is a unit thereby proving that a right Artinian ring is a quotient ring. The chain $cR \supseteq c^2R \supseteq \ldots$ stops. Therefore there exists an integer k such that $c^kR = c^{k+1}R$. Hence $c^k = c^{k+1}t$ for some $t \in R$. Because c is right regular it follows that $ct = 1$. Now $c(1-tc) = 0$. Therefore $tc = 1 = ct$.

For a commutative ring the process of forming the quotient ring is straightforward and is an easy generalisation of the way in which the field of rational numbers can be constructed from the ring of integers by forming quotients (i.e. fractions). In non-commutative rings it is sometimes easy to find the quotient ring: If R is the ring of all (upper triangular) 2 by 2 matrices over the ring of integers then an element of R is regular if and only if it has non-zero determinant and the quotient ring of R is the ring of all (upper triangular) 2 by 2 matrices over the field of rational numbers. Also it can be shown that if R is the integral group ring ZG of a finite group G then the quotient ring of R is the group algebra of G over the rational numbers. (In both these cases elements of the quotient ring have the special form $ac^{-1}$ with $a \in R$ and c a non-zero integer).

The Ore condition determines whether the quotient ring exists: We state here without proof the well known theorem that R has a right quotient ring if and only if it satisfies the *right Ore condition* with respect to the set $C(0)$ of regular elements of R, i.e. if and only if given $a, c \in R$ with $c \in C(0)$ there exist $b, d \in R$ with $d \in C(0)$ such that $ad = cb$. In general, to say that R satisfies the right Ore condition with respect to a subset S means that the above condition holds when $C(0)$ is replaced by S. Clearly a commutative ring satisfies the Ore condition with respect to any subset.

Let R be a ring with a right quotient ring Q. Suppose also that R satisfies the left Ore condition. We shall show that Q is also the left quotient ring of R. Let $q \in Q$. Then $q = ac^{-1}$ for some $a \in R$, $c \in C(0)$. By the left Ore condition there exist $b \in R$, $d \in C(0)$ such that $da = bc$. Therefore $q = ac^{-1} = d^{-1}b$.

Example 1.26. Let F be a field, x an indeterminate, and set

$$R = \begin{bmatrix} F & F[x] \\ 0 & F[x] \end{bmatrix}$$

An element of R is regular if and only if its diagonal entries are non-zero. It can be shown that R has a right quotient ring by verifying the right Ore condition or by showing directly that

$$\begin{bmatrix} F & F(x) \\ 0 & F(x) \end{bmatrix}$$

is the right quotient ring of R. We shall now show that R does not satisfy the left Ore condition with respect to $\mathcal{C}(0)$. Set $a = e_{12}$ and $c = e_{11} + xe_{22}$ then $c \in \mathcal{C}(0)$. If $r \in R$ then the (1,2)-entry of rc is divisible by x but that of ra is a constant. Suppose that b, $d \in R$ with da = bc then the (1,2)-entry of da must be 0 so that the (1,1)-entry of d is 0. Thus d is not regular, so that R does not satisfy the left Ore condition with respect to $\mathcal{C}(0)$.

Note that the integral domain constructed in 1.22 also has a right quotient ring but not a left quotient ring. □

Suppose now that R is a ring which has a right quotient ring Q. Let I be a right ideal of Q. Let $q \in I$, then $qc \in R$ for some regular element c of R. Thus $q = qcc^{-1}$ with $qc \in I \cap R$ and $c^{-1} \in Q$. It follows easily that $I = (I \cap R)Q$. This makes it easy to transfer certain conditions from R to Q. For example, if R is semi-prime, prime, simple, right Noetherian, right Artinian or has finite right Goldie dimension then the same condition holds in Q.

Another useful fact is that any finite set of elements of the right quotient ring Q of R can be written as fractions with the same denominator. We shall first show this for two elements. Let $q_1, q_2 \in Q$. Then there are regular elements $c_1$ and $c_2$ of R such that $q_1c_1 \in R$ and $q_2c_2 \in R$. In order to write $q_1$ and $q_2$ as fractions with a common denominator c we must find a regular element c of R such that $q_1c \in R$ and $q_2c \in R$. It is enough to show that $c_1R \cap c_2R$ contains a regular element c. Because $c_2$ is regular, there are elements $x_1$ and $x_2$ of R with $x_1$ regular such that $c_1x_1 = c_2x_2$. Set $c = c_1x_1$ then c is clearly regular and is an element of $c_1R \cap c_2R$. It is now easy to extend the argument to deal with any finite set of elements of Q. An easy consequence is that if I is a right ideal of R then $IQ = \{xc^{-1}: x \in I, c$ is regular in R$\}$.

<u>Theorem 1.27</u>. (Goldie) Let R be any ring, then R has a right quotient ring which is semi-simple Artinian if and only if R is a semi-prime right Goldie ring.

<u>Proof</u>. Firstly suppose that R is a semi-prime right Goldie ring. Let $a, c \in R$ with c regular. Because cR is an essential right ideal (1.11), so also is $\{r \in R: ar \in cR\}$ (1.1). Therefore there is a regular element d of R such that $ad \in cR$ (1.10). Thus R satisfies the right Ore condition and so has a right quotient ring Q.

We know that Q is semi-prime. Let A and B be right ideals of Q with $B \subseteq A$ and suppose that $B \cap R$ is an essential right R-submodule of $A \cap R$. We shall show that A = B. Let $x \in A \cap R$ then there is an essential right ideal K of R such that $xK \subseteq B \cap R$ (1.1). But K contains a regular element c (1.10). Thus $xc \in B \cap R$ and $x = xcc^{-1} \in (B \cap R)Q = B$. Hence $A \cap R \subseteq B$ so that $A = (A \cap R)Q \subseteq B$, as required. Now let $A_1, A_2, \ldots$ be a strictly decreasing chain of right ideals of Q. For each i there is a non-zero right ideal $X_i$ of R such that $X_i \subseteq R \cap A_i$ and $X_i \cap (R \cap A_{i+1}) = 0$. The sum $X_1 + X_2 + \ldots$ is direct so that there is only a finite number of the $A_i$. Therefore Q is right Artinian.

Conversely, suppose that R has a right quotient ring Q which is semi-simple

Artinian. Because Q has the a.c.c. for right annihilators so also does R. We leave it as an exercise to show that R has finite right Goldie dimension; it follows easily from the fact that if A and B are right ideals of R with $A \cap B = 0$ then $AQ \cap BQ = 0$. Let N be an ideal of R such that $N^2 = 0$. We shall show that N = 0 and it follows that R is semi-prime. There is a central idempotent element e of Q such that QNQ = eQ. We have $e = x_1n_1y_1 + \dots + x_kn_ky_k$ for some $x_i, y_i \in Q$ and $n_i \in N$. Because the $y_i$ can be written as fractions with the same denominator, there is a regular element c of R such that $ec \in QN$. Hence 0 = ecN = ceN so that eN = 0. But $N \subseteq QNQ$, i.e. $N \subseteq eQ$. Therefore N = eN = 0. □

Theorem 1.28. (Goldie) Let R be any ring, then R has a right quotient ring which is simple Artinian if and only if R is a prime right Goldie ring.

Proof. Let R be a prime right Goldie ring, then R has a semi-simple Artinian right quotient ring Q (1.27). Let I be a non-zero ideal of Q then $I \cap R$ is a non-zero ideal of R. Hence $I \cap R$ is essential as a right ideal of R and so contains a regular element c of R (1.10). But c is a unit of Q. Therefore I = Q. The proof of the converse is similar to the final argument of 1.27. □

Before leaving quotient rings we shall prove some results which will be needed later.

Lemma 1.29. Let R be a right Noetherian ring which has a right quotient ring Q. Suppose that Q is finitely-generated as a right R-module, then Q = R.

Proof. Let c be a regular element of R. Because Q is Noetherian as a right R-module, the chain $c^{-1}R \subseteq c^{-2}R \subseteq c^{-3}R \subseteq \dots$ stabilises. Therefore there is a positive integer n such that $c^{-n}R = c^{-(n+1)}R$. Hence $R = c^{-1}R$ so that $c^{-1} \in R$. Therefore Q = R. □

Lemma 1.30. (a) (Goldie) Let R be a ring with the a.c.c. for right annihilators, let S be a multiplicatively-closed set of left regular elements of R, and suppose that R satisfies the right Ore condition with respect to S, then the elements of S are regular.

(b) Let R be a ring with finite right Goldie dimension and let c and d be elements of R such that cd is right regular, then c is right regular.

Proof. (a) Let $a \in R$ and $c \in S$ with $ca = 0$. There is a positive integer n such that $r(c^n) = r(c^{n+1})$. Because $c^n \in S$ there exist $b \in R$ and $d \in S$ such that $ad = c^n b$. We have $c^{n+1}b = cad = 0$ so that $b \in r(c^{n+1})$. Therefore $0 = c^n b = ad$. Now d is left regular so that $a = 0$.

(b) Because $r(cd) = 0$ we have $dR \cap r(c) = 0$. Also $r(d) = 0$ because $r(cd) = 0$. Hence dR is an essential right ideal of R (1.11) so that $r(c) = 0$. □

The next result was proved independently by A.V. Jategaonkar and A.T. Ludgate.

Theorem 1.31. Let R be a ring which has a right Noetherian right quotient ring Q and let A be an ideal of R, then AQ is an ideal of Q.

Proof. We wish to show that $QAQ \subseteq AQ$. It is enough to show that $c^{-1}AQ \subseteq AQ$ for each regular element c of R because every element of Q is of the form $rc^{-1}$ for some $r \in R$. We have $cA \subseteq A$ so that $A \subseteq c^{-1}A$. Hence $AQ \subseteq c^{-1}AQ \subseteq c^{-2}AQ \subseteq \dots$ . Because Q is right Noetherian there is a positive integer n such that $c^{-n}AQ = c^{-(n+1)}AQ$. Multiplying on the left by $c^n$ we obtain $AQ = c^{-1}AQ$. □

In 9.11 we shall give an example of a ring R which has a right quotient ring Q and an ideal A of R such that AQ is not an ideal of Q.

## Nil Subrings

We shall show that nil subrings are nilpotent in the presence of certain chain conditions (cf. 1.8). We shall often apply this kind of result to the largest nil ideal N of a ring R, and when N is nilpotent we shall call it the *nilpotent radical* of R. The treatment we give below is taken from a paper by J.W. Fisher.

A subring N of a ring R is said to be *right* T-*nilpotent* if for each sequence $x_1, x_2, \ldots$ of elements of N there is a positive integer k such that $x_k x_{k-1} \ldots x_2 x_1 = 0$.

Lemma 1.32. Let R be a ring with the a.c.c. for right annihilators and let N be a nil subring of R which is not right T-nilpotent. Then there are elements $a_1, a_2, \ldots$ of N such that (i) each $a_i R$ is non-zero and the sum $a_1 R + a_2 R + \ldots$ is direct, and (ii) if $S_i = \{a_j : j \geq i\}$ then the $\ell(S_i)$ form a strictly increasing chain of left annihilators.

Proof. We shall say that an element x of N has an infinite chain if there is a sequence $x_1, x_2, \ldots$ of elements of N such that, for each n, the product $x_n \ldots x_2 x_1 x \neq 0$. In this case we say that $\ldots, x_2, x_1$ is a chain for x. Because N is not right T-nilpotent there is an element y of N which has an infinite chain. We define a sequence $y_1, y_2, \ldots$ of elements of N inductively as follows: Let $K_n = \{x \in N : x y_{n-1} \ldots y_2 y_1 y$ has an infinite chain$\}$ and let $y_n \in K_n$ with $r(y_n)$ as large as possible.

It is not hard to show that $r(y_i) = r(y_{i+j} \ldots y_{i+1} y_i)$ for all positive integers i and j; as an example we shall show that $r(y_1) = r(y_3 y_2 y_1)$. Clearly $r(y_1) \subseteq r(y_3 y_2 y_1)$. We have $y_1 \in K_1$. Also $y_3 \in K_3$ so that $y_3 y_2 y_1 y$ has an infinite chain. Thus $y_3 y_2 y_1 \in K_1$. By the maximality of $r(y_1)$ we have $r(y_1) = r(y_3 y_2 y_1)$.

For each n set $a_n = y_n \dots y_2 y_1 y$. Suppose that $y_1 a_n \neq 0$ for some n. We shall obtain a contradiction. For each k we have $y_k \dots y_2 y_1 a_n \neq 0$ because $r(y_1) = r(y_k \dots y_2 y_1)$. Thus $\dots, y_3, y_2$ form an infinite chain for $y_1 a_n = y_1 y_n \dots y_2 y_1 y$. Therefore $y_1 y_n \dots y_2 y_1 \in K_1$. By maximality of $r(y_1)$ we have $r(y_1) = r(y_1 y_n \dots y_2 y_1)$. This leads to a contradiction because $y_n \dots y_2 y_1$ is an element of N and hence is nilpotent; for example, if $(y_n \dots y_2 y_1)^2 = 0$ then $y_1 (y_n \dots y_2 y_1)^2 = 0$ so that $y_1 y_n \dots y_2 y_1 = 0$ and hence $y_1 a_n = 0$.

Similarly we have $y_i a_n = 0$ for all $n \geq i$. It is now straightforward to check that the $a_i$ have the desired properties. □

Lemma 1.33. Let R be a ring with the a.c.c. for right annihilators and let N be a non-nilpotent subring of R, then N is not right T-nilpotent.

Proof. There is a positive integer k such that $r(N^k) = r(N^s)$ for all $s \geq k$. Set $K = N^k$ then $r(K) = r(K^2)$. We have $K^2 \neq 0$ because N is not nilpotent. Thus there is an element $x_1$ of K such that $Kx_1 \neq 0$. Hence $K^2 x_1 \neq 0$ so that $Kx_2 x_1 \neq 0$ for some $x_2 \in K$, and so on. □

The next two results follow immediately from 1.32 and 1.33.

Theorem 1.34. (Herstein and Small) Let R be a ring with the a.c.c. for right annihilators and for left annihilators, then each nil subring of R is nilpotent.

Theorem 1.35. (Lanski) Let R be a right Goldie ring then each nil subring of R is nilpotent.

Bergman's Example

We are very grateful to G.M. Bergman for allowing us to include the following unpublished example of a prime ring of right and left Goldie dimension 1 which is not a Goldie ring. In fact the example is primitive and has certain other interesting additional features.

Example 1.36. (Bergman) In what follows, "function" means real-valued function of a real variable. Let W be the set of all surjective analytic functions f such that f has positive derivative and $f(x + 1) = f(x) + 1$ for all x, then W is a group with respect to the operation of composition of functions. We wish to construct a countable subgroup G of W such that G is dense in W with respect to the supremum norm.

Let F be the set of truncated Fourier series of period 1 with rational coefficients. Thus an element of F is of the form $f(x) = \sum_{k=0}^{n} (a_k\cos(2\pi kx) + b_k\sin(2\pi kx))$ where the $a_k$ and $b_k$ are rational. Let G be the subgroup of W generated by all elements f of W which are given by $f(x) = x + g(x)$ for all x with $g \in F$. Note that g is not an arbitrary element of F because $f'(x) > 0$ for all x. Clearly G is a countable subgroup of W. Let $w \in W$ and let $\varepsilon$ be positive. We wish to show that there exists $f \in G$ such that $|w(x) - f(x)|<\varepsilon$ for all x and, because of the periodicity-like property of elements of W, it is enough to have this inequality for all $x \in [0,1]$. We can write $w(x) = x + u(x)$ where u is a periodic analytic function of period 1. We have $u'(x) > -1$ for all x. Because u' is continuous, there is a positive number $\delta$ such that $u'(x) > \delta -1$ for all x. By standard results of analysis there exists $g \in F$ such that $|u(x) - g(x)| < \varepsilon$ and $|u'(x) - g'(x)| < \delta$ for all x. Set $f(x) = x + g(x)$ for all x then $f \in G$ and $|w(x) - f(x)| < \varepsilon$ for all x.

Let u and v be distinct elements of W then, because u - v is analytic, the set of x such that $u(x) = v(x)$ is discrete and hence countable. Therefore there is a real number p such that no two distinct elements of G agree on p.

Let f, $g \in W$. We write $f \geq g$ if and only if $f(p) \geq g(p)$. If $f \geq g$ then

$hf \geq hg$ for all $h \in W$. Let $c(x) = x + 1$ for all $x$ then $c$ is a central element of $W$. Let $S = \{g \in G: g \geq 1\}$ and $T = \{s \in S: c > s\}$ then $S$ is a sub-semigroup of $G$. Note that $c$ is a central element of $S$ and $c \notin T$. Let $s \in S$ then $s \geq 1$ so that $cs \geq c$, i.e. $cs \notin T$. On the other hand, let $s \in S$ with $s \notin T$ then $s \geq c$ so that $c^{-1}s \in S$, i.e. $s \in cS$. Therefore $S$ is the union of the disjoint sets $cS$ and $T$.

We can now define the ring we want. Let $K$ be a field then $c$ is a central element of the semigroup algebra $KS$. Set $R = KS/cKS$ then we can regard $R$ as being the $K$-algebra which has the elements of $T$ as a $K$-basis. Note that if $t_1, t_2 \in T$ with $t_1t_2 \geq c$ (i.e. $t_1t_2(p) \geq p + 1$) then in $R$ we have $t_1t_2 = 0$.

Let $I$ be a non-zero right ideal of $R$. We shall show that $I$ contains an element of $T$. Let $x = a_1t_1 + \ldots + a_nt_n$ be a non-zero element of $I$ where each $a_i$ is a non-zero element of $K$ and the $t_i$ are distinct elements of $T$. We may suppose that $1 \geq t_1^{-1} > t_2^{-1} > \ldots > t_n^{-1} \geq c^{-1}$. We have $t_it_2^{-1}c \geq c$ for all $i \neq 1$ and $t_2^{-1}c \in T$. Multiplying $x$ on the right by $t_2^{-1}c$ we have $a_1t_1t_2^{-1}c \in I$ with $t_1t_2^{-1}c \in T$. Also if $u, v \in T$ then $v \geq u$ if and only if $u^{-1}v \in T$, i.e. if and only if $vR \subseteq uR$. It follows that if $I$ and $J$ are non-zero right ideals of $R$ then $tR \subseteq I \cap J$ for some $t \in T$, so that $R$ has right Goldie dimension 1.

Let $T^*$ be the set of non-identity elements of $T$ and let $U$ be the $K$-subspace of $R$ generated by $T^*$, then clearly $U$ is a maximal ideal of $R$. Let $g, h \in T^*$. Thus $p < g(p) < p + 1$ and $p < h(p) < p + 1$. Because $G$ is dense in $W$, there exists $e \in G$ such that $e(p) > p$ and $e(g(p)) < h(p)$; in fact $e \in T^*$. Thus $eg < h$, so that $1 < g^{-1}e^{-1}h$. Set $f = g^{-1}e^{-1}h$ then $h = egf$ with $e, f \in T^*$. It follows that $U \subseteq U^3$ so that $U = U^2$. Also if $I$ is any proper ideal of $R$ then $g \in I$ for some $g \in T^*$. From the relationship $h = egf$ obtained above it follows that $T^* \subseteq I$ and hence that $I = U$. Thus $U$ is idempotent and is the only proper ideal of R. Therefore R is prime.

Let u be a non-zero element of R with $u^2 = 0$, e.g. $u \in T^*$ with $u(x) = x + \frac{1}{2}$ for all x, then uR is an essential right ideal of R and u.uR = 0. Thus the right singular ideal Z(R) of R is non-zero so that Z(R) = U. To show that R has neither the a.c.c. nor the d.c.c. for right annihilators we note that if $t \in T^*$ then $tR = r(ct^{-1})$ (cf. 1.6).

We shall now show that R is right primitive. Because G is dense in W there exists $g \in G$ such that $p < g(p) < g(p + \frac{1}{2}) \leq p + \frac{1}{2}$. Thus $g \in T^*$ and $g^n(p) \leq p + \frac{1}{2}$ for each positive integer n. Therefore g is not a nilpotent element of R, from which it follows that 1 + g is not a unit of R. Let M be a maximal right ideal of R which contains 1 + g then $g \notin M$ so that U is not contained in M. Therefore R/M is a faithful simple right R-module.

Thus, in particular, R is a right primitive ring of right Goldie dimension 1. The corresponding left-handed properties can be proved similarly, basically by replacing the ordering $f \geq g$ used above by the ordering defined by $f \geq g$ if and only if $g^{-1}(p) \geq f^{-1}(p)$ (in this case $f \geq g$ implies $fh \geq gh$).□

Kerr's Example.

It is easy to show that if R has finite right Goldie dimension k then $M_n(R)$ has finite dimension kn. On the other hand, if R has the a.c.c. for right annihilators it does not follow that $M_n(R)$ has the a.c.c. for right annihilators if $n \neq 1$. In fact there is an example due to J.C. Shepherdson of an integral domain D such that $M_2(D)$ does not satisfy the annihilator conditions. Thus if R is a right Goldie ring then $M_n(R)$ has finite right Goldie dimension, but it was not known until recently whether $M_n(R)$ always has the a.c.c. for right annihilators in these circumstances. This question was settled by J.W. Kerr and we are very grateful to her for allowing us to include her example of a commutative Goldie ring R such that $M_2(R)$ is not a a Goldie ring.

Lemma 1.37. Let F and G be free right R-modules with bases $X = \{x_i\}$ and $Y = \{y_j\}$ respectively and let $f:F \to G$ be a homomorphism such that $f(X) = Y$. For each $y_j \in Y$ choose $z_j \in X$ such that $f(z_j) = y_j$. Then Ker(f) is a free module with basis $\bigcup_j \{z_j - x_i : x_i \in X,\ x_i \neq z_j,\ f(x_i) = f(z_j)\}$.

Proof. Straightforward. □

Lemma 1.38. Let R be a ring with sequences of elements $s_1, s_2, \ldots$ and $t_1, t_2, \ldots$ such that $s_i t_j = 0$ if and only if $i \neq j$, then R does not satisfy the a.c.c. for right annihilators.

Proof. For each i set $S_i = \{s_j : j \geq i\}$ then $r(S_i) \subseteq r(S_{i+1})$. Also $t_i \in r(S_{i+1})$ and $t_i \notin r(S_i)$ (because $s_i t_i \neq 0$). Therefore $r(S_i) \neq r(S_{i+1})$. □

Example 1.39. (Kerr) We shall construct a commutative Goldie ring R such that $M_2(R)$ is not a Goldie ring. The strategy is to construct a commutative Goldie ring R with elements $u_i, v_i, x_i, y_i$ for $i = 1,2,\ldots$ such that $u_i x_j = v_i y_j$ if and only if $i \neq j$. Working in $M_2(R)$ set $s_i = u_i e_{11} - v_i e_{12}$ and $t_i = x_i e_{11} + y_i e_{21}$, then $s_i t_j = 0$ if and only if $i \neq j$ so that $M_2(R)$ does not have the a.c.c. for right annihilators (1.38).

Let K be any commutative integral domain and let S be the commutative polynomial ring over K in four sequences of indeterminates $a_i$, $b_i$, $c_i$ and $d_i$ $(i = 1,2,\ldots)$. Let T be the commutative polynomial ring over K in indeterminates $e_i$, $f_i$ and g $(i = 1,2,\ldots)$; note that g is a single indeterminate and is not a mis-print for $g_i$. Let $h:S \to T$ be the K-algebra homomorphism determined by the rules $h(a_i) = ge_i$, $h(b_i) = e_i$, $h(c_i) = f_i$, $h(d_i) = gf_i$ for all i. Set $z_{ij} = a_i c_j - b_i d_j$ then $h(z_{ij}) = 0$. Set P = Ker(h) then P is a prime ideal of S because T is an integral domain. We have $z_{ij} \in P$ for all i and j. It is easy to show that P is homogeneous, i.e. that if $s \in S$ then $s \in P$ if and only if each homogeneous component of s belongs to P. Set $P_i = \{p \in P: p \text{ is homogeneous of degree } i\}$ then $P_i$ is a K-submodule of P and $P = P_0 \oplus P_1 \oplus P_2 \oplus \ldots$ It is easy to show that $P_0 = P_1 = 0$.

Let F be the free K-submodule of S with the monomials of degree 2 in S as basis and let f be the restriction of h to F, then $\text{Ker}(f) = P_2$. Because f maps monomials to monomials, it follows from 1.37 that $P_2$ is the free K-module with basis $B = \{z_{ij}: \text{all } i \text{ and } j\} \cup \{a_ib_j - a_jb_i, c_id_j - c_jd_i: i < j\}$. Let Q be the ideal of S consisting of all elements of S with zero constant term, and set $J = P_3 + P_4 + \ldots$ then J is an ideal of S and $QP \subseteq J$. Let I be the K-submodule of P generated by J together with $z_{ii} - z_{jj}$ for all i and j and all elements of B except those of the form $z_{ii}$. Because $QP \subseteq J$ and $S = K + Q$ it is easy to show that I is an ideal of S. Also $P = Kz_{11} \oplus I$ as K-modules so that $P/I \cong K$.

Let * denote image in the ring $S^* = S/I$ ($S^*$ is the ring we want). We have $P^* \cong K$ so that $P^*$ is a uniform ideal of $S^*$. Suppose that $kp \in I$ for some $k \in K$ and $p \in P$. Because $P = Kz_{11} \oplus I$ we have $k = 0$ or $p \in I$. Hence Q is the largest ideal of S such that $QP \subseteq I$, so that $Q^* = r(P^*)$ (for convenience we shall use the notation of right or left annihilator even though $S^*$ is commutative). Also $Q^* = r(U)$ for each non-zero ideal U of $S^*$ with $U \subseteq P^*$. Because $P^*$ is prime and does not contain $Q^*$ we have $P^* = \ell(Q^*)$. Let V and W be non-zero ideals of $S^*$ with $W = r(V)$ and $V = \ell(W)$ then $VW \subseteq P^*$ and so we may suppose without loss of generality that $V \subseteq P^*$. Thus $VQ^* = 0$ from which it follows that $W = Q^*$ and $V = P^*$. Hence $P^*$ and $Q^*$ are the only non-zero annihilator ideals in $S^*$. It is similarly easy to show that $S^*$ has Goldie dimension 2.

Set $R = S^*$, $u_i = a_i^*$, $v_i = b_i^*$, $x_i = c_i^*$, and $y_i = d_i^*$, then R is a commutative Goldie ring. We have $u_i x_j = v_i v_j = z_{ij}^*$, and we know that $z_{ij}^* = 0$ if and only if $i \neq j$ (by definition of I). Therefore R satisfies the conditions laid down in the first paragraph of 1.39. □

Remarks

(1) Many of the result of this chapter, and Goldie's theorem in particular, are true for rings without identity element.

(2) The following is the sketch of a proof of 1.34 which is more direct than the proof given above but which does not give any information about Goldie rings. We temporarily drop the assumption that rings have identity element. Let R be a nil ring with the a.c.c. and d.c.c. for right annihilators. We shall show that R is nilpotent. Let P be maximal among all ideals of R of the form $r(I)$ where I is a non-zero ideal of R, then R/P is a prime nil ring with the a.c.c. for right annihilators. Hence P = R (1.7). Thus $\ell(R) \neq 0$. There is a positive integer n such that $\ell(R^n) = \ell(R^{n+1})$. If $\ell(R^n) \neq R$ then the proof above that $\ell(R) \neq 0$ can be applied in $R/\ell(R^n)$ to show that there exists $x \in R$ with $xR \subseteq \ell(R^n)$ and $x \notin \ell(R^n)$, i.e. $xR^{n+1} = 0$ and $xR^n \neq 0$, which is a contradiction. Therefore $\ell(R^n) = R$ so that $R^{n+1} = 0$.

(3) For further results and examples concerning nil subrings see the relevant papers by J.W. Fisher, C. Lanski, and I.N. Herstein and L.W. Small.

(4) It was shown by R. Gordon, T.H. Lenagan, and J.C. Robson that if R is a ring with right Krull dimension then nil subrings of R are nilpotent (see, for example, [100]).

(5) Examples with similar properties to those of 1.36 can be found in [6].

(6) A prime P.I. ring is a Goldie ring by Posner's theorem [114] and so has a quotient ring, but G.M. Bergman has given an example of a semi-prime P.I. ring which has no quotient ring [2].

(7) J.W. Fisher has shown that the singular ideals of a semi-prime P.I. ring are zero [46], and I.N. Herstein and L.W. Small have shown that an element of such a ring is right regular if and only if it is left regular [80].

(8) J.W. Kerr has also constructed examples of the following kinds: A commutative Goldie ring such that R[x] is not a Goldie ring, a commutative ring with a.c.c. for annihilator ideals but with no bound on the lengths of chains of annihilators (this example can be taken to be nilpotent).

# 2 The rank of a module

In 1964 A.W. Goldie introduced the notion of a rank for a suitable kind of module which reduces to the usual notion of rank for an Abelian group or dimension of a vector space. In the last few years this idea has been used to prove several new results and to give new proofs of some known ones. In these notes we shall give several of these applications, for example, in chapter 3 we shall use this concept of rank to prove a non-commutative generalisation of the principal ideal theorem of commutative algebra. In this chapter we establish the basic properties of the rank function and give one quick application (to Small's theorem). The rank function we are about to define is often called the reduced rank if there is possibility of confusing it with some other rank.

To avoid complicated conditions we shall deal with a finitely generated right R-module M over a right Noetherian ring R. However, it is worth pointing out that what we are about to do will work under weaker assumptions. For M we need only assume, for example, that all factor modules of M have finite Goldie dimension. This condition is satisfied by any module with Krull dimension. Also we only need the nil radical N of R to be nilpotent and R/N to be a right Goldie ring. We shall see later examples of rings of this type which are not right Noetherian.

We begin with the special case where R is a semi-prime right Noetherian ring. Let M be a finitely generated right R-module. We know that Goldie's theorem applies to R and so R satisfies the right Ore condition with respect to the set $C(0)$ of regular elements of R. This makes it possible to talk about torsion elements of M in a way which is similar to the notion of torsion elements of an Abelian group. We call an element x of M a *torsion element* if $xc = 0$ for some $c \in C(0)$. Let T(M) denote the set of all torsion elements of M. Then T(M) is a submodule of M; for example let $x \in T(M)$ and $r \in R$. We have $xc = 0$ for some $c \in C(0)$. Also $rd = cs$ for some $d \in C(0)$ and $s \in R$. Thus $xrd = 0$ so that $xr \in T(M)$. In fact in this case T(M) coincides with the singular submodule of M because a right ideal of R is essential if and only if it contains an element of C(0). We define $\rho(M)$ the *rank* of M to be the Goldie dimension of M/T(M). We shall not assume that the reader is familiar with tensor products, but those who are will probably recognise that $\rho(M)$ is the length of $M \otimes_R Q$ as a right Q-module where Q is the right quotient ring of R.

We denote by dim K the Goldie dimension of a module K. We shall now show that if R is a semi-prime right Noetherian ring and X is a finitely generated right R-module with a submodule Y such that X/Y is torsion-free then

$$\dim(X) = \dim(Y) + \dim(X/Y). \qquad (*)$$

There is a submodule Y' of X such that $Y \cap Y' = 0$ and $Y \oplus Y'$ is essential in X. Let $x \in X$ and $x \notin Y$. There is an essential right ideal E of R such that $xE \subseteq Y \oplus Y'$ (1.1). But E contains a regular element c by Goldie's theorem. Thus $xc \in Y \oplus Y'$ and $xc \notin Y$ because X/Y is torsion-free. It follows that if Z is a submodule of X such that $Z \cap (Y \oplus Y') = Y$ then $Z = Y$, so that $(Y \oplus Y')/Y$ is an essential submodule of X/Y. Therefore,

$$\dim(X/Y) = \dim((Y \oplus Y')/Y) = \dim(Y') .$$

But it follows from one of the remarks after 1.9 that

$$\dim(Y \oplus Y') = \dim(Y) + \dim(Y').$$

Thus $\dim(X) = \dim(Y \oplus Y') = \dim(Y) + \dim(X/Y)$ proving (*). □

We now show that the rank function is additive.

Lemma 2.1. Let R be a semi-prime right Noetherian ring and let M be a finitely generated right R-module. Let K be a submodule of M then

$$\rho(M) = \rho(K) + \rho(M/K).$$

Proof. Using tensor products the result is easy because tensoring with the right quotient ring of R is an exact functor and the length of a module is additive on short exact sequences. As an alternative the following is an elementary direct proof.

Let L be the submodule of M such that $K \subseteq L$ and $L/K = T(M/K)$.

We shall first show that $K + T(M)$ is essential in L. Let $y \in L$. Then $yd \in K$ for some regular element d. If $yd \neq 0$ then yd is a non-zero element of $yR \cap [K + T(M)]$. If $yd = 0$ then $y \in T(M)$. Thus $K + T(M)$ is essential in L. Hence $\dim(L) = \dim(K + T(M))$. (i)

Now $M/L \cong (M/K)/(L/K) = (M/K)/T(M/K)$ so that M/L is torsion-free. Therefore

$$\begin{aligned}\rho(M/K) &= \dim(M/L) = \dim(M) - \dim(L) \text{ by (*) above}\\ &= \dim(M) - \dim(T(M)) - [\dim(L) - \dim(T(M))] \qquad \text{(ii)}\end{aligned}$$

But

$$\begin{aligned}\rho(K) &= \dim(K/T(K)) = \dim(K/(K \cap T(M))\\ &= \dim((K + T(M))/T(M))\\ &= \dim(K + T(M)) - \dim(T(M)) \text{ by (*) since } (K + T(M))/T(M)) \text{ is a}\end{aligned}$$

torsion free module. Therefore $\rho(K) = \dim L - \dim(T(M))$ by (i). Thus now (ii) yields

$$\begin{aligned}\rho(M/K) &= \dim(M) - \dim(T(M)) - \rho(K)\\ &= \dim(M/T(M)) - \rho(K) \text{ by (*) since } M/T(M) \text{ is torsion-free. Thus}\end{aligned}$$

$\rho(M) = \rho(M/K) + \rho(K)$. □

We now extend the definition of $\rho$ to finitely generated modules over a right Noetherian ring which is not necessarily semi-prime. Let M be a finitely generated module over a right Noetherian ring R. Let N be the nilpotent radical of R. Then there exists an integer $k \geq 1$ such that $MN^k = 0$. Also R/N is a semi-prime ring. We define $\rho(M)$ the *rank* of M to be $\rho(M) = \sum_{i=0}^{k-1} \rho(MN^i/MN^{i+1})$ $(N^0 = R)$, where on the right hand side $MN^i/MN^{i+1}$ is viewed as a right R/N-module and $\rho(MN^i/MN^{i+1})$ is calculated over the semi-prime ring R/N.

Note that $\rho(M) = \rho(MN) + \rho(M/MN)$.

It is easily seen that isomorphic right R-modules have the same rank.

Once again the rank function is additive. The proof we give is an adaptation by T.H. Lenagan of the Schrier refinement argument.

Theorem 2.2. Let M be a finitely generated right module over a right Noetherian ring R. Let N be the nilpotent radical of R.

(a) If K is a submodule of M then $\rho(M) = \rho(K) + \rho(M/K)$.

(b) $\rho(M) = 0$ if and only if M is torsion with respect to $C(N)$ i.e. if and only if for each $x \in M$ there is an element c of R such that $c + N$ is a regular element of R/N and $xc = 0$.

(c) If $\rho(M) = 0$ and $x_1, \ldots, x_n \in M$ then there exists $c \in C(N)$ such that $x_i c = 0$ for all i.

Proof. (a) There is an integer $k \geq 1$ such that $MN^k = 0$. We prove the result by induction on k. If $k = 1$ we have the result by 2.1. Now assume the result for all finitely generated modules X such that $XN^s = 0$; $s \leq k - 1$.

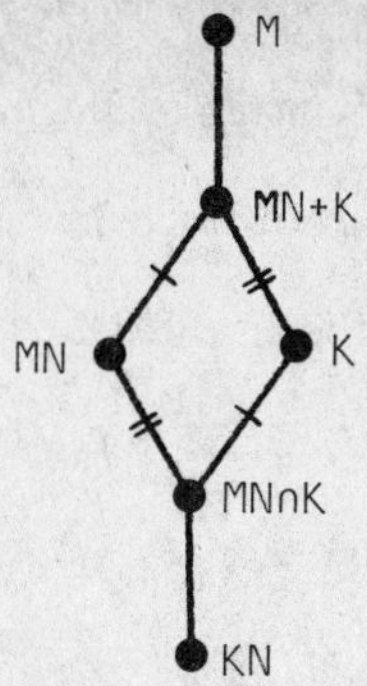


Let K be a submodule of M. The factors corresponding to the opposite sides of the diamond in the diagram are isomorphic and hence have the same rank.

We have $\rho(M) = \rho(MN) + \rho(M/MN)$ by definition.

Now, $\rho(MN) = \rho(MN \cap K) + \rho(MN/(MN \cap K))$

$$= \rho(KN) + \rho((MN \cap K)/KN) + \rho(MN/(MN \cap K)) \qquad \text{(i)}$$

by induction hypothesis on MN.

$$= \rho(KN) + \rho((MN \cap K)/KN) + \rho((MN + K)/K) \qquad \text{(ii)}$$

and $\rho(M/MN) = \rho((MN + K)/MN) + \rho((M/MN)/(MN + K)/MN)$ by 2.1

$$= \rho((MN + K)/MN) + \rho(M/(MN + K))$$

$$= \rho(K/(MN \cap K)) + \rho(M/(MN + K)) . \qquad \text{(iii)}$$

From (i), (ii) and (iii) we have

$$\rho(M) = \rho(KN) + [\rho((MN \cap K)/KN) + \rho(K/(MN \cap K))] + [\rho((MN + K)/K) + \rho(M/(MN + K)]$$

$$= \rho(KN) + \rho(K/KN) + \rho(M/K)$$ using 2.1 and the definition of $\rho(M/K)$.

Thus $\rho(M) = \rho(K) + \rho(M/K)$ by definition of $\rho(K)$.

(b) If X is a finitely-generated right R/N-module then $\rho(X) = 0$ if and only if $X/T(X) = 0$, i.e. if and only if X is a torsion module over R/N. The desired result follows from this and induction on the smallest positive integer k such that $MN^k = 0$.

(c) As in (b) the general case reduces by induction to the case where $MN = 0$, i.e. without loss of generality we may suppose that R is semi-prime. We have $x_1, \ldots, x_n \in M$ and $\rho(M) = 0$. Therefore for each i there is a regular element $c_i$ of R such that $x_i c_i = 0$. Each $c_i R$ is an essential right ideal of R and hence so also is $c_1 R \cap \ldots \cap c_n R$. Therefore, by Goldie's theorem, there is a regular element c of R such that $c \in c_1 R \cap \ldots \cap c_n R$. We have $x_i c = 0$ for all i. □

As the first of several applications of the rank function in these notes we give new proofs of some well-known results concerning right Noetherian rings.

Theorem 2.3. Let R be a right Noetherian ring and let N denote the nilpotent radical of R.

(a) (Djabali) The right regular elements of R are regular mod(N).

(b) (Goldie) Let $a, c \in R$ with c right regular then there exist $b \in R$ and $d \in \mathcal{C}(N)$ such that $ad = cb$.

(c) (Small) R has a right Artinian right quotient ring if and only if $\mathcal{C}(0) = \mathcal{C}(N)$.

Proof. Let c be a right regular element of R, then left multiplication by c gives a right R-module isomorphism between R and cR. Therefore $\rho(R) = \rho(cR)$ so that $\rho(R/cR) = 0$. Thus R/cR is torsion with respect to $\mathcal{C}(N)$, i.e. for each $a \in R$ there exists $d \in \mathcal{C}(N)$ such that $ad \in cR$; this proves (b). In particular there is an element x of $\mathcal{C}(N)$ such that $x \in cR$. Therefore $c \in \mathcal{C}(N)$ by 1.30(b) applied to the ring R/N. This establishes (a).

(c) Suppose that $\mathcal{C}(0) = \mathcal{C}(N)$. Let $a \in R$ and $c \in \mathcal{C}(0)$ then, by (b), there exist $b \in R$ and $d \in \mathcal{C}(N)$ such that $ad = cb$. But $d \in \mathcal{C}(0)$. Thus R satisfies the right Ore condition with respect to $\mathcal{C}(0)$ and hence R has a right quotient ring Q. Let A and B be right ideals of Q with B strictly contained in A then $(A \cap R)/(B \cap R)$ is a non-zero module which is torsion-free with respect to $C(0)$. But $C(0) = C(N)$. Therefore $\rho((A \cap R)/(B \cap R))$ is positive (2.2(b)). It now follows easily that Q is right Artinian because R does not have an infinite descending chain of right ideals such that the corresponding factor modules all have positive rank. We leave the proof of the converse of (c) as an exercise (cf. 9.2). □

Remarks

(1) Let R be a semi-prime right Noetherian ring and let X and Y be finitely-generated right R-modules with torsion submodules T(X) and T(Y) respectively. The Goldie dimension of a module is defined in terms of direct sums of uniform submodules and therefore can be expressed in terms of sums and intersections of submodules. Therefore $\rho(X) = \rho(Y)$ whenever there is a one-to-one correspondence between the submodules of X/T(X) and those of Y/T(Y) which preserves sums and intersections. We shall make use of this observation in Chapter 3.

(2) The idea of using the rank function to prove the existence of a right quotient ring has also been used by Warfield in [156] (we shall give one of these applications in the chapter on serial rings), and by Lenagan in [100] to show that if R is a ring with right Krull dimension then R has a right Artinian right quotient ring if and only if $C(0) = C(N)$.

(3) In general, the left and right ranks of a bimodule are unrelated. Set $F = Z/2Z$,

$$R = \begin{bmatrix} Z & F \\ 0 & F \end{bmatrix} \quad \text{and} \quad N = \begin{bmatrix} 0 & F \\ 0 & 0 \end{bmatrix} ,$$

then R is a left and right Noetherian ring and N is the nilpotent radical of R. Because $R/N \cong Z \oplus F$, the set $C(N)$ consists of all elements of R both of whose diagonal entries are non-zero. Thus $2e_{11} + e_{22} \in C(N)$ and $(2e_{11} + e_{22})N = 0$, so that the rank of N as a left R-module is 0. On the other hand, then rank of N as a right R-module is equal to the rank N as a right R/N-module because $N^2 = 0$, and this coincides with the rank of F as a right F-module which is 1.

(4) It is an immediate consequence of Small's theorem (2.3(c)) that if R is a left and right Noetherian ring then R has a right Artinian right quotient ring if and only if R has a left Artinian left quotient ring.

# 3 The invertible ideal theorem

The main aim of this chapter is to prove a generalisation to non-commutative rings of the principal ideal theorem of commutative algebra, and the proof we shall give will use the rank function defined in Chapter 2. We shall begin by recalling the principal ideal theorem and its extension to invertible ideals in the commutative case; we shall then discuss possible ways of formulating non-commutative generalisations before proving our main result: Let R be a right Noetherian ring with an invertible ideal $X \neq R$ and let P be a prime ideal of R which is minimal over X, then P has rank at most 1 (the terminology will be explained as we go along).

Let R be a commutative Noetherian ring, let a be a non-unit of R, and let P be a prime ideal minimal over aR (i.e. P is a prime ideal which is minimal amongst all prime ideals which contain aR). The principal ideal theorem asserts that P has rank at most 1, i.e. there do not exist prime ideals $P_1$ and $P_2$ of R such that $P_2 \subsetneq P_1 \subsetneq P$. Another way of stating this is to assume that R is a commutative Noetherian integral domain with a prime ideal P which is minimal over aR where a is a non-zero non-unit of R, then P does not properly contain any non-zero prime ideal of R. In the non-commutative case aR is not in general an ideal, but that does not matter because we can regard P as being minimal amongst primes containing a. The following example shows that the most natural candidate for a non-commutative version of the principal ideal theorem is false.

<u>Example 3.1</u>. (Procesi) Let $R = M_2(Z[x])$ where Z is the ring of integers and x is an indeterminate. Then R is a left and right Noetherian prime ring.

Set $a = 2e_{11} + xe_{22}$ and $P = RaR$, then a is a non-zero non-unit of R and $P = M_2(2Z + xZ[x])$. Thus P is a prime ideal of R which is clearly minimal over a. However, P properly contains the non-zero prime ideal 2R of R. □

Thus some restriction must be placed on the element a in the non-commutative case if the principal ideal theorem is to work for prime ideals minimal over a. We shall return to this problem later. However we formulate a non-commutative principal ideal theorem, the usual proofs of the commutative case cannot easily be modified because they use localisation and this technique is not available in the non-commutative case (at least, not in any form which would be useful here).

An ideal X of a ring R is said to be *invertible* if the following conditions are satisfied: There is an over-ring S of R (often the quotient ring of R) such that if $A = \{s \in S: sX \subseteq R\}$ and $B = \{s \in S: Xs \subseteq R\}$ then $AX = XB = R$. In this situation we have $A = B$ and we write $X^{-1} = A = B$. If R is a commutative integral domain then every non-zero principal ideal of R is invertible (in the quotient field of R). Every non-zero ideal of a Dedekind domain is invertible, and this shows that invertible ideals need not be principal.

<u>Example 3.2</u>. Let

$$R = \begin{bmatrix} Z & 2Z \\ Z & Z \end{bmatrix} \quad \text{and} \quad X = \begin{bmatrix} 2Z & 2Z \\ Z & 2Z \end{bmatrix}$$

then X is an invertible ideal of R. To justify this claim, let S be the quotient ring of R, i.e. $S = M_2(Q)$, and set

$$X^{-1} = \begin{bmatrix} Z & Z \\ \frac{1}{2}Z & Z \end{bmatrix}$$

then $X^{-1}$ is a subset of S and $X^{-1}X = XX^{-1} = R$. □

Let $R$ be a commutative Noetherian integral domain with an invertible ideal $X \neq R$. Let $P$ be a prime minimal over $X$ and let $T$ be the localisation of $R$ at $P$. Because $X$ is invertible in some over-ring $S$ of $R$, there are elements $s_1,\ldots,s_n$ of $S$ and $x_1,\ldots,x_n$ of $X$ such that $1 = s_1x_1 + \ldots + s_nx_n$ and $s_iX \subseteq R$ for all $i$. Let $f_i:X \to R$ be defined by $f_i(x) = s_ix$ for all $x \in X$, then each $f_i$ is an $R$-module homomorphism and $\{x_i,f_i\}$ is a dual basis for $X$. Thus $X$ is a projective $R$-module so that $XT$ is a projective ideal of $T$. But $T$ is a local ring. Hence $XT$ is a free $T$-module and so is a principal ideal of $T$. But $PT$ is a prime minimal over $XT$, so that $PT$ (and hence also $P$) has rank at most 1, by the principal ideal theorem applied in $T$ to the principal ideal $XT$. Thus it is straightforward, using standard commutative ring theory, to use the principal ideal theorem to prove the "invertible ideal theorem". In the non-commutative case we do not have the technique of localisation available, and we shall in fact first prove an invertible ideal theorem and then derive from it a version of the principal ideal theorem.

Let $X$ be an ideal of a ring $R$. Suppose that $X$ is invertible in an over-ring $S$. Let $A$, $B$ be right $R$-submodules of $S$. Then

$$\begin{aligned}(A \cap B)X &\subseteq AX \cap BX = (AX \cap BX)R = (AX \cap BX)X^{-1}X \\ &\subseteq (AXX^{-1} \cap BXX^{-1})X = (AR \cap BR)X = (A \cap B)X.\end{aligned}$$

Thus $(A \cap B)X = AX \cap BX$. Similarly $(A \cap B)X^{-1} = AX^{-1} \cap BX^{-1}$.

<u>Lemma 3.3</u>. (Ginn) Let $X$ be an invertible ideal of a right Noetherian ring $R$. Then $X$ has the right Artin-Rees property, i.e., if $I$ is a right ideal of $R$ then there is a positive integer $n$ such that $I \cap X^n \subseteq IX$.

Proof. We have $XX^{-1} = X^{-1}X = R$ where $X^{-1}$ is defined in some over-ring of $R$. Set $Y = X^{-1}$. We have $R \subseteq Y$ so that $Y = YR \subseteq Y^2 \subseteq Y^3 \subseteq \dots$. Thus $IY \cap R \subseteq IY^2 \cap R \subseteq IY^3 \cap R \subseteq \dots$ is an ascending chain of right ideals of $R$. Therefore there is a positive integer $n-1$ such that $IY^{n-1} \cap R = IY^n \cap R$. Hence $(IY^{n-1} \cap R)X^n = (IY^n \cap R)X^n$. It follows that $IX \cap X^n = I \cap X^n$. Thus $I \cap X^n \subseteq IX$. □

Theorem 3.4. (Chatters, Goldie, Hajarnavis, Lenagan) Let $R$ be a right Noetherian ring, let $X$ be an invertible ideal of $R$ with $X \neq R$, and let $P$ be a prime ideal minimal over $X$, then $P$ has rank at most 1.

Proof. The first step is to reduce the problem to the case where $R$ is a prime ring. We have $X^{-1}X = XX^{-1} = R$ where $X^{-1}$ is defined in an over-ring $S$ of $R$. Set $Y = X^{-1}$. Let $Q$ be a prime ideal of $R$ which does not contain $X$; we shall show that $X \cap Q = XQ = QX$ and that $(X + Q)/Q$ is an invertible ideal of $R/Q$. Set $I = R \cap YQ$. Then $I$ is an ideal of $R$ and $XI = X(R \cap YQ) = XR \cap XYQ = X \cap Q \subseteq Q$. Therefore $I \subseteq Q$, so that $X \cap Q = XI \subseteq XQ$. Hence $X \cap Q = XQ$, and similarly we also have $X \cap Q = QX$.

Note that $Y^n \subseteq S$ for all positive integers $n$. Set $T = \bigcup_{n=1}^{\infty} Y^n$. Then $T$ is a subring of $S$ which contains $R$. We shall show that $QT$ is an ideal of $T$. Clearly it is enough to show that $TQ \subseteq QT$. Let $t \in T$ and $q \in Q$. Then there is a positive integer $n$ such that $t \in Y^n$. Because $XQ = QX$ we have $QY = YXQY = YQXY = YQ$. Thus $tq \in Y^nQ = QY^n$ so that $tq \in QT$ as required. Let $r \in QT \cap R$. Then $r \in QY^i$ for some positive integer $i$. We have $rX^i \subseteq Q$ so that $r \in Q$. Hence $QT \cap R = Q$. Thus $R/Q = R/(QT \cap R) \cong (R + QT)/QT$. Let $f$ be the natural homomorphism from $T$ to $T/QT$ then $f(X)f(Y) = f(Y)f(X) = f(R) = (R + QT)/QT$, so that $(X + QT)/QT$ is an invertible ideal of the ring $(R + QT)/QT$. Therefore $(X + Q)/Q$ is an invertible ideal of $R/Q$.

Suppose that there are prime ideals $P_1$ and $P_2$ of R such that $P_2 \subsetneq P_1 \subsetneq P$; we shall obtain a contradiction. $(X + P_2)/P_2$ is an invertible ideal of $R/P_2$ because $P_2$ does not contain X. Also $P/P_2$ is minimal over $(X + P_2)/P_2$. We may therefore assume without loss of generality that $P_2 = 0$.

We have thus achieved the aim of the first part of the proof which is to reduce to the case where R is a prime ring. Thus we assume that R is a prime right Noetherian ring with an invertible ideal $X \neq R$, the prime ideal P is minimal over X and $P_1$ is a non-zero prime ideal of R with $P_1 \subsetneq P$. We shall prove the theorem by establishing a contradiction. Because $P_1$ is a non-zero ideal of a prime right Noetherian ring, $P_1$ contains a regular element y, by Goldie's theorem. There is a positive integer n such that $yR \cap X^n \subseteq yX$ (3.3). We have $yR \cap X^{n+1} = yR \cap X^n \cap X^{n+1} \subseteq yX \cap X^{n+1} = yRX \cap X^nX = (yR \cap X^n)X \subseteq yX^2$. Repeating this process we eventually obtain $yR \cap X^{2n} \subseteq yX^{n+1} \subseteq yX^n$. Since P is minimal over the invertible ideal $X^n$, we may without loss of generality assume that n = 2, i.e. that $yR \cap X^2 \subseteq yX$. But $yR \cap (X^2 + yX) = (yR \cap X^2) + yX$, by the modular law. Therefore $yR \cap (X^2 + yX) = yX$.

Let $\rho(M)$ denote the rank of a finitely-generated right R/X-module M. We have

$$\begin{aligned} \rho((X^2 + yR)/(X^2 + yX)) &= \rho((X^2 + yX + yR)/(X^2 + yX)) \\ &= \rho(yR/(yR \cap (X^2 + yX))) \\ &= \rho(yR/yX) \\ &= \rho(R/X) \qquad \text{(i)} \end{aligned}$$

The last equality holds because left multiplication by the regular element y gives an isomorphism between R/X and yR/yX. We shall prove shortly that if A and B are right ideals of R with $AX \subseteq B \subseteq A$ then $\rho(A/B) = \rho(AX/BX)$.

Taking $A = R$ and $B = X + yR$ we obtain

$$\rho(R/(X + yR)) = \rho(X/(X^2 + yX)) \tag{ii}$$

Therefore

$$\begin{aligned}\rho((X + yR)/(X^2 + yX)) &= \rho(X/(X^2 + yX)) + \rho((X + yR)/X) \\ &= \rho(R/(X + yR)) + \rho((X + yR)/X) \text{ by (ii)} \\ &= \rho(R/X) \qquad\qquad (iii)\end{aligned}$$

Here we have used the additivity of the rank function (2.2(a)). We have $X^2 + yX \subseteq X^2 + yR \subseteq X + yR$ and (i) and (iii) yield $\rho((X + yR)/(X^2 + yX)) = \rho((X^2 + yR)/(X^2 + yX))$. Therefore $\rho((X + yR)/(X^2 + yR)) = 0$. Let $N$ be the ideal of $R$ such that $X \subseteq N$ and $N/X$ is the nilpotent radical of $R/X$. By Lemma 2.2(b), $(X + yR)/(X^2 + yR)$ is a torsion module with respect to the set of all elements of $R/X$ which are regular mod$(N/X)$, i.e. $(X + yR)/(X^2+yR)$ is torsion with respect to $C(N)$. Thus for each $x \in X$ there is an element $d$ of $C(N)$ such that $xd \in X^2 + yR$.

Because $X$ is invertible there are elements $x_1,\ldots,x_n$ of $X$ and elements $s_1,\ldots,s_n$ of $X^{-1}$ such that $1 = s_1x_1 + \ldots + s_nx_n$. For each $x \in X$ we have $x = xs_1x_1 + \ldots + xs_nx_n$ with $xs_i \in R$ for each $i$. Thus $X$ is finitely-generated as a left ideal of $R$ by elements $x_1,\ldots,x_n$. Now there is an element $c$ of $C(N)$ such that $x_ic \in X^2 + yR$ for all $i$ (2.2(c)). Thus $x_ic \in X^2 + P_1$ for all $i$. Since $X^2 + P_1$ is an ideal of $R$, we have $Rx_ic \subseteq X^2 + P_1$ for all $i$, so that $Xc \subseteq X^2 + P_1$. Hence $Xc \subseteq X^2 + (X \cap P_1)$. But $P_1$ is a prime ideal which does not contain $X$, and it was shown earlier in the proof that in these circumstances we have $X \cap P_1 = XP_1$. Therefore $Xc \subseteq X^2 + XP_1$ and multiplying on the left by $X^{-1}$ we obtain $Rc \subseteq X + P_1$. But $X \subseteq P$ and $P_1 \subseteq P$ so that $c \in P$. This is a contradiction (1.25).

To complete the proof of Theorem 3.4 we must justify the assertion made earlier. We claimed that if A and B are right ideals of the prime right Noetherian ring R such that $AX \subseteq B \subseteq A$ where X is an invertible ideal of R then $\rho(A/B) = \rho(AX/BX)$. Here we are regarding A/B and AX/BX as right R/X-modules. Recall that N is the ideal of R such that $X \subseteq N$ and N/X is the nilpotent radical of R/X. We have $XNX^{-1} \subseteq XX^{-1} = R$ so that $XNX^{-1}$ is an ideal of R. There is a positive integer i such that $N^i \subseteq X$. We have $(XNX^{-1})^i = XN^iX^{-1} \subseteq X^2X^{-1} = X$ so that $XNX^{-1} \subseteq N$. Thus $XN \subseteq NX$, and by symmetry we have $XN = NX$. In order to show that $\rho(A/B) = \rho(AX/BX)$ it is enough to show that $\rho((B + AN^j)/(B + AN^{j+1})) = \rho((BX + AXN^j)/(BX + AXN^{j+1}))$ for each integer $j \geq 0$. But $(BX + AXN^j)/(BX + AXN^{j+1}) = (B+AN^j)X/(B+AN^{j+1})X$. Therefore without loss of generality we may suppose that $AN \subseteq B$ and $AXN \subseteq BX$.

Thus we can consider A/B and AX/BX to be finitely-generated right R/N-modules. If C and D are right ideals of R with $B \subseteq C \subseteq A$ and $BX \subseteq D \subseteq AX$ then $B \subseteq DX^{-1} \subseteq A$ and $BX \subseteq CX \subseteq AX$. The mappings $C \to CX$ and $D \to DX^{-1}$ induce a one-to-one correspondence between the submodules of A/B and those of AX/BX which preserves sums and intersections. In order to show that $\rho(A/B) = \rho(AX/BX)$ it is thus enough to show that the torsion submodules of A/B and AX/BX correspond in the above way (cf. Remark (1) at the end of Chapter 2). Let C/B and D/BX be the torsion submodules of A/B and AX/BX respectively. We must show that $CX \subseteq D$ and $DX^{-1} \subseteq C$, i.e. that CX/BX and $DX^{-1}/B$ are torsion R/N-modules.

Let Y be either X or $X^{-1}$. We must show that if $W \subseteq V \subseteq U$ are suitable right R-modules with $UN \subseteq W$ and V/W is torsion with respect to $\mathcal{C}(N)$ then VY/WY is also torsion. Without loss of generality we may suppose that $W = 0$. Let $a \in VY$ and set $K = \{r \in R: ar = 0\}$. We have $VYN = VNY \subseteq UNY$ so that $VYN = 0$. Thus $N \subseteq K$. We wish to show that K contains an element of

$C(N)$. By Goldie's theorem this is equivalent to showing that $K/N$ is an essential right ideal of $R/N$. Let $I$ be a right ideal of $R$ such that $I \cap K = N$. Thus $IY^{-1} \cap KY^{-1} = NY^{-1}$. Let $s \in IY^{-1}$ then $s \in Y^{-1}$. We have $as \in VYY^{-1}$, i.e. $as \in V$. But $V$ is torsion with respect to $C(N)$, so there exists $c \in C(N)$ such that $asc = 0$. Hence $ascY = 0$. But $scY \subseteq sY \subseteq Y^{-1}Y = R$. Therefore $scY \subseteq K$. Also $sc \in IY^{-1}c$ so that $sc \in IY^{-1}$. Hence $scY \subseteq I$. Therefore $scY \subseteq I \cap K$ so that $scY \subseteq N$. Thus $sc \in NY^{-1}$. Now $NY^{-1} = Y^{-1}N$ so that $Ysc \subseteq N$. But $Ys \subseteq YIY^{-1} \subseteq YY^{-1} = R$ and $c \in C(N)$ with $Ysc \subseteq N$. Therefore $Ys \subseteq N$ so that $Rs \subseteq Y^{-1}N$. But $Y^{-1}N = NY^{-1}$ so that $s \in NY^{-1}$. Thus we have proved that $IY^{-1} \subseteq NY^{-1}$ so that $I \subseteq N$ and hence $I = N$. In other words, if $I$ is a right ideal of $R$ such that $N \subseteq I$ and $(K/N) \cap (I/N) = 0$ then $I/N = 0$. Therefore $K/N$ is an essential right ideal of $R/N$, as required.□

An element $x$ of a ring $R$ is said to be a *normalising element* of $R$ if $xR = Rx$. Units of $R$ and central elements of $R$ are normalising elements, but there are normalising elements which are neither units nor central, e.g. the element $2(e_{12} + e_{21})$ of $M_2(Z)$ or the element $e_{12}$ of the ring of 2 by 2 upper triangular matrices over $Z$.

Corollary 3.5. (Jategaonkar) Let $R$ be a right Noetherian ring, let $x$ be a normalising element of $R$ which is not a unit, and let $P$ be a prime ideal minimal over $xR$, then $P$ has rank at most 1.

Proof. We shall in fact give two proofs. In the first we obtain this result as a direct consequence of Theorem 3.4; in the second we shall show how the proof of Theorem 3.4 can be considerably simplified in this special case.

First Proof. Suppose that $P_1$ is a prime ideal of $R$ which is strictly contained in $P$. It is enough to show that there are no prime ideals of $R$

strictly between $P_1$ and $P$, and therefore without loss of generality we may suppose that $P_1 = 0$. Thus to complete the first proof we need only show that if $x$ is a non-zero normalising element of a prime right Noetherian ring $R$ then $xR$ is an invertible ideal of $R$. Because $R$ is prime and $xR = Rx$ so that $xR$ is a non-zero two-sided ideal of $R$, we have $0 = r(Rx) = r(x)$ and $0 = \ell(xR) = \ell(x)$. Thus $x$ is a regular element of $R$.

Let $S$ be the right quotient ring of $R$ and set $X = xR$, $A = \{s \in S : sX \subseteq R\}$, and $B = \{s \in S : Xs \subseteq R\}$. Let $a \in A$ then $aX \subseteq R$, so that $ax \in R$. Thus $A \subseteq Rx^{-1}$ and the reverse inclusion is obvious. Thus $A = Rx^{-1}$ and similarly $B = x^{-1}R$. We have $AX = Rx^{-1}xR = R = XB$ so that $X$ is invertible in $S$. □

Second Proof. As in the first proof we can easily reduce to the case where $R$ is prime and $x$ is a regular element of $R$ such that $xR = Rx$. Set $X = xR$. The most difficult part of the proof of Theorem 3.4 was to show that $\rho(R/(X + yR)) = \rho(X/(X^2 + yX))$, which in this case becomes $\rho(R/(xR + yR)) = \rho(xR/(x^2R+yxR))$. It is enough to show that $\rho(xR/(x^2R+yxR)) = \rho(xR/(x^2R+xyR))$ because $xR/(x^2R + xyR) \cong R/(xR + yR)$. We shall show how to prove this in a moment, but before doing so we note that the rest of the proof of Theorem 3.4 goes through either without alteration or with obvious simplifications;. for example, it is clear that $X$ is finitely-generated (in fact principal) as a left ideal of $R$.

The main idea is to take advantage of the fact that $x$ is a regular normalising element of $R$ in order to define an inner automorphism of $R$. Define $f:R \to R$ by $f(r) = xrx^{-1}$ for all $r \in R$. Then $f$ is an automorphism of $R$ and $f(x^2R) = f(xRx) = x^2Rxx^{-1} = x^2R$. Set $R' = R/x^2R$ and $r' = r + x^2R$ for all $r \in R$. Then $f$ induces an automorphism $g$ of $R'$ given by $g(r') = f(r)'$. We wish to show that $\rho(xR/(x^2R + yxR)) = \rho(xR/(x^2R + xyR))$, i.e. that

$\rho(x'R'/y'x'R') = \rho(x'R'/x'y'R')$. But $g(x'R') = x'R'$ and $g(y'x'R') = x'y'R'$.

To simplify the notation, we can sum up as follows: We have a ring S with right ideals $B \subseteq A$ and an automorphism h of S; we wish to show that $\rho(A/B) = \rho(h(A)/h(B))$ where $\rho(M)$ denotes the rank of an R/xR-module M. Let N be the ideal of R such that $xR \subseteq N$ and N/xR is the nilpotent radical of R/xR. In order to prove that $\rho(A/B) = \rho(h(A)/h(B))$ we may without loss of generality suppose that $AN \subseteq B$. Although h is not in general a module homomorphism, it is clear that h induces a one-to-one correspondence between the submodules of A/B and those of h(A)/h(B) which preserves sums and intersections. In view of Remark (1) at the end of Chapter 2, it is enough to show that the torsion submodules of A/B and h(A)/h(B) correspond in this way. But this follows easily from the fact that $f(C(N)) = C(N)$. □

The proofs we have just given of Corollary 3.5 can easily be modified to cover the case where $xR = Ry$ for some $y \in R$ instead of requiring that $xR = Rx$.

Theorem 3.6. (Goldie) Let R be a prime right Noetherian ring which satisfies a polynomial identity and let x be a regular non-unit of R. Let B be the sum of all the two-sided ideals of R which are contained in xR and let P be a prime ideal minimal over B. Suppose that x is not regular mod(P). Then P has rank at most 1.

Proof. We will only indicate how the proof of Theorem 3.4 can be modified to prove this result. We replace X by xR almost everywhere except that we take ranks of R/B-modules rather than R/X-modules and we take N to be the ideal of R such that $B \subseteq N$ and N/B is the nilpotent radical of R/B. We suppose that $P_1$ is a non-zero prime ideal of R such that $P_1$ is strictly contained in P; we shall obtain a contradiction. Let y be a non-zero central element of R which belongs to $P_1$. Because R/yR is a right Noetherian ring there is a positive integer n such that $x^n r \in yR$ whenever $x^{n+1} r \in yR$. Without loss of generality we can take n = 1 and obtain $yR \cap (x^2R + yxR) = yxR$. Following the argument used before it is easy to show that $\rho((x^2R + yR)/(x^2R+yxR))= \rho(R/xR)$. The proof that $\rho((xR + yR)/(x^2R + yxR)) = \rho(R/xR)$ is much easier than the corresponding part of the proof of Theorem 3.4 because $\rho(xR/(x^2R + yxR)) = \rho(xR/(x^2R + xyR)) = \rho(R/(xR + yR))$. (Recall that y is central). Thus $\rho((xR + yR)/(x^2R + yR)) = 0$, so that $xc \in x^2R + yR$ for some $c \in C(N)$. Hence $xc \in x^2R + P_1$. But $B \subseteq xR$ and B is not contained in $P_1$. Therefore by 1.30(b) $x \in C(P_1)$. But $xc - x^2R \in P_1$ for some $r \in R$. Therefore $c - xr \in P_1$, so that $c \in xR + P_1$. But $c \in C(N)$ and $C(N) \subseteq C(P)$ (because P is minimal over N). Therefore $c + p_1 \in C(P)$ for all $p_1 \in P_1$. But there exists $p_1 \in P_1$ such that $c - p_1 = xr$. Thus $xr \in C(P)$. It follows by 1.30(b) that $x \in C(P)$. □

The following example shows that it is possible to have $x \in C(P)$ in the situation of Theorem 3.6.

<u>Example 3.7</u>. Let

$$R = \begin{bmatrix} Z & 2Z \\ Z & Z \end{bmatrix} \quad \text{and} \quad P = \begin{bmatrix} 2Z & 2Z \\ Z & Z \end{bmatrix}$$

then R is a left and right Noetherian prime P.I. ring and P is a prime ideal of R. Let $x = e_{11} + 2e_{22}$ then x is a regular non-unit of R. The largest two-sided ideal of R which is contained in xR is $M_2(2Z)$; thus we have $B = M_2(2Z)$ in the notation of Theorem 3.6. The prime P is minimal over B and $x \in C(P)$, but nevertheless P has rank 1. □

Remarks

(1) It is not known whether Theorem 3.6 is true in the case where x is regular mod(P).

(2) There are several versions of the principal ideal theorem for non-commutative rings in the literature but it was two papers by Jategaonkar [85 and 86] which inspired the material of this chapter.

(3) We will use the invertible ideal theorem in Chapter 12 when we consider rings of finite global dimension.

(4) Let R be a left and right Noetherian hereditary prime ring and assume that R is right bounded (i.e. that every essential right ideal of R contains a non-zero two-sided ideal). It was shown by T.H. Lenagan that R has enough invertible ideals, i.e. that each non-zero ideal of R contains an invertible ideal of R [96].

# 4 The Artinian radical

In Chapters 4 and 5 we shall discuss what can be thought of as the Artinian part of a Noetherian ring. In this chapter we shall define the Artinian radical and establish its basic properties, and in the next chapter we shall give some applications of these ideas.

Let R be a right Noetherian ring with nilpotent radical N such that the ring R/N is right Artinian. Then the modules $N^k/N^{k+1}$ are finitely generated right R/N-modules, and hence are Artinian. Therefore R is right Artinian. Since a semi-prime ring is left Artinian if and only if it is right Artinian, it follows that a left Artinian right Noetherian ring is right Artinian.

Theorem 4.1. (Lenagan) Let R be a right Noetherian ring and let I be an ideal of R which has finite length as a left R-module, then I is Artinian as a right R-module.

Proof. Because R is right Noetherian, each non-zero ideal A of R contains an ideal B such that A/B is a minimal ideal of R/B. Together with the fact that I is Artinian as a left R-module, this implies that there are ideals $I_1,\ldots,I_n$ of R such that $0 = I_n \subseteq I_{n-1} \subseteq \cdots \subseteq I_1 = I$ and $I_j/I_{j+1}$ is a minimal ideal of $R/I_{j+1}$. To prove that I is Artinian as a right R-module it is enough to show that each $I_j/I_{j+1}$ is Artinian. Without loss of generality we may therefore suppose that I is a minimal ideal of R.

Set $P = r(I)$ then it is easy to show that P is a prime ideal of R. Set $S = R/P$ and let $Z(I)$ be the torsion submodule of I as a right S-module, i.e. $Z(I) = \{x \in I: xc = 0 \text{ for some } c \in C(P)\}$. Because Z(I) is an ideal of R we have either $Z(I) = 0$ or $Z(I) = I$. Suppose that $Z(I) = I$; we shall obtain a contradiction. There are elements $x_1,\ldots,x_k$ of I such that $I = Rx_1+\ldots+Rx_k$ (because I is Noetherian as a left R-module). For each i there exists $c_i \in C(P)$ such that $x_i c_i = 0$. Each $(c_i R + P)/P$ is an essential right ideal of S, and so, by Goldie's theorem, there exists $c \in C(P)$ such that $c \in c_i R + P$ for all i. We have $Ic = 0$. Hence $c \in r(I)$, i.e. $c \in P$, which is a contradiction.

Thus I is torsion-free as a right S-module. Let $d \in C(P)$ then $I \supseteq Id \supseteq Id^2 \supseteq \ldots$ . But I is Artinian as a left R-module. Therefore there is a positive integer t such that $Id^t = Id^{t+1}$. For each $x \in I$ we have $xd^t = yd^{t+1}$ for some $y \in I$, and, because I is torsion-free with respect to C(P), this gives $x = yd$. Thus $I = Id$.

Therefore we can in a natural way make I into a right Q-module where Q is the right quotient ring of S. But Q is a simple Artinian ring (1.28). Thus there is a simple right Q-module M such that every non-zero right Q-module is the direct sum of a set of copies of M. In particular, I is a direct sum of copies of M. Therefore M is finitely-generated as a right S-module. It follows that Q also is finitely-generated as a right S-module, so that $Q = S$ (1.29). Thus I is finitely-generated as a right module over the simple Artinian ring S, so that I is Artinian as a right S-module and hence also as a right R-module. □

<u>Corollary 4.2</u>. (Lenagan) Let R be a left and right Noetherian ring then an ideal of R is Artinian as a right R-module if and only if it is Artinian as a left R-module.

<u>Corollary 4.3</u>. Let R be a right Noetherian ring and let I be an ideal of R which has finite length as a left R-module, then there is an ideal X of R

such that R/X is right Artinian and IX = 0.

Proof. As in the proof of 4.1 there are ideals $I_1,\dots,I_n$ of R such that $0 = I_n \subseteq \dots \subseteq I_1 = I$ and $I_j/I_{j+1}$ is a minimal ideal of $R/I_{j+1}$. Also as in 4.1 there is an ideal $P_j$ such that $R/P_j$ is simple Artinian and $I_jP_j \subseteq I_{j+1}$. Set $X = P_1P_2\dots P_{n-1}$ then X has the required properties. □

Corollary 4.4. Let R be a left and right Noetherian ring with ideals I and X such that R/X is Artinian, then there is an ideal Y of R such that R/Y is Artinian and $YI \subseteq IX$.

Proof. Set S = R/IX then I/IX is a finitely-generated right R/X-module and hence is Artinian as a right S-module. There is an ideal L of S such that S/L is Artinian and L(I/IX) = 0 (4.3). Let Y be the ideal of R such that $IX \subseteq Y$ and Y/IX = L then Y has the required properties. □

Before defining the Artinian radical we shall prove a theorem of S.M.Ginn and P.B. Moss which gives a way of proving that certain Noetherian rings are Artinian, and in order to do that we need the following lemma. If M is a right R-module and S is a non-empty subset of M we shall write

$$r(S) = \{r \in R: sr = 0 \text{ for all } s \in S\}.$$

Lemma 4.5. (a) Let M be a right R-module and suppose that there are elements $x_1,\dots,x_n$ of M such that $r(M) = r(x_1) \cap \dots \cap r(x_n)$. Then $R/r(M)$ is isomorphic to a right R-submodule of $x_1R \oplus \dots \oplus x_nR$.

(b) Let I be a right ideal of a left Noetherian ring R then there is a finite subset S of I such that $r(I) = r(S)$.

(c) Let R be a left and right Noetherian ring and let I be a right ideal of R, then I is Artinian as a right R-module if and only if $R/r(I)$ is an Artinian ring.

Proof. (a) Define $f: R \to x_1R \oplus \ldots \oplus x_nR$ by $f(r) = (x_1r,\ldots,x_nr)$ for all $r \in R$, then f is a right R-module homomorphism and $\mathrm{Ker}(f) = r(x_1)\cap\ldots\cap r(x_n) = r(M)$.

(b) There are elements $x_1,\ldots,x_n$ of R such that $RI = Rx_1 + \ldots + Rx_n$. For each i there are finitely-many elements $r_{ij}$ of R and $y_{ij}$ of I such that $x_i = \sum_j r_{ij}y_{ij}$. Let $S = \{y_{ij}\}$ then $r(I) = r(S)$.

(c) Suppose that R is left and right Noetherian and that the right ideal I of R is Artinian as a right R-module. By (b), there are elements $x_1,\ldots,x_n$ of I such that $r(I) = r(x_1) \cap \ldots \cap r(x_n)$. Hence, by (a), the R-module $R/r(I)$ embeds in $x_1R \oplus \ldots \oplus x_nR$. But for each i we have $x_iR \subseteq I$ so that $x_iR$ is Artinian. Therefore $R/r(I)$ is Artinian as a right R-module and so is an Artinian ring. The converse of (c) is trivial. □

Theorem 4.6. (Ginn and Moss) Let R be a left and right Noetherian ring and assume that the right socle E of R is essential as a right ideal or as a left ideal, then R is Artinian.

Proof. By 4.2, E is Artinian as a left R-module. Thus if E is essential as a left ideal then so also is the left socle of R. Therefore without loss of generality we may assume that E is essential as a right ideal of R. Thus $\ell(E)$ is contained in the right singular ideal of R, so that $\ell(E) \subseteq N$ where N is the nilpotent radical of R (1.6). But E is Artinian as a left R-module so that $R/\ell(E)$ is Artinian (4.5(c)). Therefore R/N is Artinian and hence R is Artinian. □

Corollary 4.7. Let R be a left and right Noetherian ring and let I be a right ideal of R. Assume that I has an essential Artinian submodule, then I is Artinian as a right R-module.

Proof. By 4.5, $R/r(I)$ embeds in the direct sum of a finite number of copies of I. Together with the fact that I has an essential Artinian submodule, this implies that $R/r(I)$ has an essential Artinian submodule as a right R-module. Hence the right socle of $R/r(I)$ is essential as a right ideal. Therefore $R/r(I)$ is Artinian (4.6), so that I is Artinian. □

Now we can define the Artinian radical and establish its basic properties. Let R be any ring. We define the *Artinian radical* A(R) of R to be the sum of all the Artinian right ideals of R (by an Artinian right ideal we mean a right ideal which is Artinian as a right R-module). We have A(R) = R if and only if R is right Artinian, and A(R) = 0 if and only if R has no non-zero Artinian right ideals. When R is left and right Noetherian A(R) has certain symmetrical properties. We shall show, for example, that A(R) is also the sum of all the Artinian left ideals of R. The following example shows that this is not the case for a ring which is only right Noetherian.

Example 4.8. Let

$$R = \begin{bmatrix} Z & Q \\ 0 & Q \end{bmatrix}$$

then R is right but not left Noetherian. The right ideals $e_{12}Q$ and $e_{22}Q$ of R are Artinian (in fact minimal) and $R/(e_{12}Q + e_{22}Q) \cong Z$, from which it follows that $A(R) = e_{12}Q + e_{22}Q$. On the other hand, R has no non-zero Artinian left ideals (this is basically because Q has no non-zero Artinian Z-submodules). In this example A(R) coincides with the right socle of R, and A(R) is essential as a right ideal. This shows that 4.6 is not true for rings which are only right Noetherian. □

Theorem 4.9. Let R be a left and right Noetherian ring and let A(R) be the Artinian radical of R, i.e. A(R) is the sum of all the Artinian right ideals of R. Then

(a) A(R) is an ideal of R, A(R) is the unique largest Artinian right ideal of R, and A(R) is also the unique largest Artinian left ideal of R.

(b) A(R/A(R)) = 0, i.e. R/A(R) has no non-zero Artinian one-sided ideals.

(c) A(R) is a *complement* right ideal and a complement left ideal, i.e. A(R) has no proper essential extensions in R as either a right R-module or a left R-module.

(d) A(R) is a right annihilator and a left annihilator in R.

Proof. (a) If S and T are Artinian right ideals of R then so also are S + T and rS for any $r \in R$. It follows that A(R) is an ideal of R and, because R is right Noetherian, A(R) is the unique largest Artinian right ideal of R. It follows from 4.2 that A(R) is contained in the sum of all the Artinian left ideals of R. Equality follows by symmetry.

(b) Let I be a right ideal of R such that $A(R) \subseteq I$ and I/A(R) is an Artinian right ideal of R/A(R). Because A(R) is right Artinian so also is I, so that $I \subseteq A(R)$. Therefore A(R/A(R)) = 0.

(c) Let K be a right ideal of R which contains A(R) as an essential submodule. Thus K has an essential Artinian submodule so that K is Artinian (4.7). Therefore K = A(R).

(d) For convenience we write A instead of A(R). By 4.5(c), $R/r(A)$ is an Artinian ring. But $\ell(r(A))$ is a finitely-generated right $R/r(A)$-module. Therefore $\ell(r(A))$ is an Artinian right ideal, so that $\ell(r(A)) \subseteq A$. Hence $A = \ell(r(A))$. □

One of the aims of the next few results is to establish a connection between the Artinian radicals of R and R/N where N is the nilpotent radical of R.

<u>Lemma 4.10</u>. Let R be a left and right Noetherian ring and assume that P is a minimal prime ideal of R such that R/P is Artinian, then A(R) is not contained in P.

<u>Proof</u>. Let $T$ denote the set of all ideals I of R such that R/I is Artinian, then $P \in T$. Because R is Noetherian, every ideal of R contains a product of prime ideals. Thus there are minimal prime ideals $P_1,\dots,P_k$ (not necessarily distinct) such that $P_1P_2\dots P_k = 0$. We have $P = P_i$ for some i, so that $P_i \in T$ for at least one i. If $P_i \in T$ for all i then R is Artinian, i.e. A(R) = R. Suppose now that there exists j such that $P_j \notin T$. If B and C are ideals of R with $C \in T$ then there exists $D \in T$ such that $DB \subseteq BC$ (4.4). For example, if $P_2 \in T$ and $P_1 \notin T$ then there exists $Q_1 \in T$ such that $Q_1P_1 \subseteq P_1P_2$, so that $Q_1P_1P_3\dots P_k = 0$. In this way each $P_j$ which does not belong to $T$ can, in effect, be moved to the right in the product $P_1P_2\dots P_k$ until we obtain $0 = XY$ where X is a product of elements of $T$ and Y is a product of minimal primes which do not belong to $T$. Thus $X \in T$ and Y is not contained in P. Because XY = 0 and $X \in T$ we have $Y \subseteq A(R)$. □

<u>Theorem 4.11</u>. Let R be a left and right Noetherian ring. Then $A(R) \subseteq N$ if and only if A(R/N) = 0.

<u>Proof</u>. If A(R) is not nilpotent then (A(R) + N)/N is a non-zero Artinian ideal of R/N. Therefore $A(R/N) \neq 0$.

Conversely, suppose that $A(R/N) \neq 0$. There is a minimal prime ideal P of R such that A(R/N) is not contained in P/N. Let B be the ideal of R such that $N \subseteq B$ and B/N = A(R/N), then (B + P)/P is a non-zero Artinian ideal of R/P. Therefore R/P is Artinian (1.24), so that A(R) is not contained in P

(4.10). Thus A(R) is not contained in N. □

The following example shows that it is not true that A(R) = 0 if and only if A(R/N) = 0.

Example 4.12. Let F = Z/2Z and

$$R = \begin{bmatrix} Z & F \\ 0 & Z \end{bmatrix},$$

then R is left and right Noetherian and N is the set of strictly upper triangular elements of R. We have $R/N \cong Z \oplus Z$, A(R/N) = 0, and A(R) = N. □

Theorem 4.13. Let R be a left and right Noetherian ring and let A denote the Artinian radical of R, then $A + r(A)$ contains an element of C(N).

Proof. Firstly let S be a semi-prime right Goldie ring and let I be an ideal of S which is not contained in any minimal prime of S. Let K be a right ideal of S such that $I \cap K = 0$. Then KI = 0. Therefore K is contained in each minimal prime of S so that K = 0. Thus I is essential as a right ideal of S and hence contains a regular element of S, by Goldie's theorem.

Thus it is enough to prove that $A + r(A)$ is not contained in any minimal prime ideal P of R. If P does not contain $r(A)$ then P does not contain $A + r(A)$. Because A is an Artinian right ideal, $R/r(A)$ is an Artinian ring (4.5(c)). If P contains $r(A)$ then R/P is Artinian, so that P does not contain A (4.10). □

It is not true in general that $A + r(A)$ contains a regular element (4.12).

We shall now give two quick applications of 4.13, the first of which will be re-derived in two other ways later.

Theorem 4.14. (Ginn and Moss) Let R be a left and right Noetherian ring which has an Artinian quotient ring, then A(R) is a direct summand of R.

Proof. Set A = A(R). There is an element $c \in C(N)$ such that $c = a + x$ for some $a \in A$ and $x \in r(A)$ (4.13). Because R has an Artinian quotient ring, c is a regular element of R by Small's theorem (2.3(c)). We have A = Ac because A is Artinian as a left ideal. Thus $a = ec$ for some $e \in A$. We have $c = a + x = ec + x$ and $ex = 0$ so that $ec = e^2c$. Thus $e = e^2$. For each $b \in A$ we have $bc = b(ec + x) = bec$ so that $b = be$. Thus A = Re. Similarly there is an idempotent element f of A such that A = fR. It is easy to show that e = f and that e is central in R. □

Corollary 4.15. Let R be an indecomposable left and right Noetherian ring which has an Artinian quotient ring then either R is Artinian or R has no non-zero Artinian one-sided ideals.

Lemma 4.16. Let K be a nilpotent ideal of a ring R. Then $\ell(K)$ is essential as a right ideal of R.

Proof. Let I be a non-zero right ideal of R. Defining $K^0$ to be R, there exists an integer n such that $IK^n \neq 0$ but $IK^{n+1} = 0$. Thus $IK^n \subseteq I \cap \ell(K)$ and so $I \cap \ell(K) \neq 0$. □

Theorem 4.17. (Ginn and Moss) Let R be a left and right Noetherian ring whose right singular ideal is zero. Let E denote the right socle of R and suppose that E has non-zero intersection with each non-zero ideal of R, then R is Artinian.

Proof. Let A denote the Artinian radical of R. By 4.13, $A + r(A)$ contains an element c of $C(N)$. From the fact that $(cR + N)/N$ is an essential right ideal of $R/N$ it follows easily that $cR + N$ is an essential right ideal of R. Because R is right non-singular we have $\ell(cR + N) = 0$. But $cR+N \subseteq A+r(A) + N$. Therefore $0 = \ell(A + r(A) + N) = \ell(A) \cap \ell(r(A)) \cap \ell(N) = \ell(A) \cap A \cap \ell(N)$ (4.9(d)). Hence $A \cap \ell(A) = 0$ (4.16). But $E \subseteq A$ so that $E \cap \ell(A) = 0$. Therefore $\ell(A) = 0$, so that $A = r(\ell(A)) = R$ (4.9(d)). Therefore R is Artinian. □

The assumption that R is right non-singular cannot be dropped from 4.17, i.e. it is possible to have a right and left Noetherian ring R which is not Artinian but whose Artinian radical has non-zero intersection with every non-zero ideal of R. The first example of this kind was constructed by W.Borho and L.W. Small, but the following example due to K.A. Brown is easier to explain.

Example 4.18. (Brown) Let C be the field of complex numbers and let U be the ring of polynomials over C in indeterminates x, y, z subject to the relations $xy - yx = z$, $xz - zx = -2x$, and $yz - zy = 2y$. In fact U is the universal enveloping algebra of the Lie algebra of 2 by 2 complex matrices with zero trace, so that U is a left and right Noetherian integral domain. Set $q = 4xy + z^2 - 2z$ then it is easy to show that q is a central element of U. Let K be the set of elements of U with zero constant term, then K is an idempotent ideal of U and $U/K \cong C$. Also K is the only proper ideal of U which strictly contains qU. The ring U/qU is an integral domain and is not a division ring (because K/qU is a proper ideal of U/qU). Thus U/qU is not Artinian. For justifications of the statements made in this paragraph see 1.8, 2.8.8, 4.9.22 and 8.44 of algèbre enveloppontes by J. Dixmier.

Set $R = U/qK$, $P = qU/qK$, and $M = K/qK$. Then R has the following properties: R is left and right Noetherian but not Artinian; the only proper ideals of R are P and M; $P \neq 0$, $P^2 = 0$, P is generated by a central element, and R/P is a domain; $M = M^2$ and $R/M \cong C$; $PM = MP = 0$; P is a one-dimensional vector space over C and P is minimal as a right ideal and as a left ideal. Clearly P is the Artinian radical of R and every non-zero ideal of R contains P. Suppose that R has a right quotient ring Q, then MQ is an ideal of Q (1.31), and $MQ \neq Q$ because $PM = 0$. We shall show in a moment that $C(0) = C(M)$, and it follows easily that elements of Q which do not belong to MQ are units of Q. Therefore MQ is the Jacobson radical of Q, which contradicts the fact that $(MQ)^2 = MQ$ (Nakayama's lemma, 10.1).

Therefore R does not have a right quotient ring, and similarly R does not have a left quotient ring. Clearly $C(0) \subseteq C(M)$ because $PM = 0$. Let $c \in C(M)$ then $c \in C(P)$ because R/P is an integral domain. Thus $\ell(c) \subseteq P$ so that $\ell(c)M = 0$. Also R/M is a field so that $R = cR + M$. Hence $\ell(c)R = 0$ so that $\ell(c) = 0$. Similarly $r(c) = 0$ and it follows that $C(0) = C(M)$. □

# 5 Applications of the Artinian radical

The theme which links the material of this chapter is the use of the Artinian radical to prove new results, or to give new proofs of known results, in the theory of Noetherian rings. The applications are to Noetherian orders in Artinian rings (extending 4.14), the decomposition theorem for Noetherian p.p. rings, the Jacobson conjecture for Noetherian rings of Krull dimension 1, and injective right ideals of Noetherian rings.

We begin by considering Noetherian orders in Artinian rings. It was shown in 4.14 that if A is the Artinian radical of such a ring R then A is a direct summand of R. We shall now establish connections between A and the socle of R/N and the intersection of all right ideals of R of the form cR with c regular, In what follows we shall not use 4.14.

Theorem 5.1. Let R be a left and right Noetherian ring which has an Artinian quotient ring, then there is a central idempotent element e of R such that

(i) (Ginn and Moss) eR is the Artinian radical of R,

(ii) (eR + N)/N is the socle of R/N, and

(iii) $eR = \cap cR = \cap Rc$ where c ranges over the regular elements of R.

Proof. Set $R^* = R/N$ and let $r^* = r + N$ for all $r \in R$. Let S be the socle of $R^*$, then $S = wR^*$ for some central idempotent element w of $R^*$ (1.23). There is an idempotent element e of R such that $e^* = w$ (cf. Remark (1) at the end of this chapter). Let A denote the Artinian radical of R and set $G = \cap cR$ and $H = \cap Rc$ where c ranges over the regular elements of R. Because A is Artinian we have $A = cA = Ac$ for each regular element c of R (as shown in the proof of 4.1). Therefore $A \subseteq G \cap H$.

We shall now show that H is an ideal of R and that H is divisible as a right R-module (i.e. $H = Hc$ for each regular element c of R). Clearly H is a left ideal of R. Let $h \in H$ and $r \in R$. Let c be a regular element of R, then because R satisfies the left Ore condition we have $xc = dr$ for some $x \in R$ and regular element d of R. But $h \in H$ so that $h = yd$ for some $y \in R$. We have $hr = ydr = yxc$ so that $hr \in Rc$. But c is an arbitrary regular element of R. Thus $hr \in H$ and H is an ideal. Now let u and v be arbitrary regular elements of R. Then $h = zu$ for some $z \in R$. Because $h \in Rvu$ we have $z \in Rv$. Hence $z \in H$ and $H = Hu$.

It will follow from 5.2 that H is Artinian as a right R-module and that $H = He$, so that $H \subseteq Re \cap A$. Hence $A = H \subseteq Re$. We shall now show that $e \in H$. It is enough to show that $e \in Rc$ where c is an arbitrary regular element of R. Set $M = R/Rc$. If $r \in R$ then $dr \in Rc$ for some regular element d of R. Thus the left R-module M is torsion with respect to $C(0)$. Hence, in order to show that $e \in H$, it is enough to show that $eX = 0$ where X is any left R-module which is torsion with respect to $C(0)$. Without loss of generality we may assume that $NX = 0$. Thus X is, in a natural way, a left $R^*$-module. Let $x \in X$ then $dx = 0$ for some regular element d of R. But $d^*$ is a regular element of $R^*$ (2.3(a)). Also the socle S of $R^*$ is Artinian so that $S = Sd^*$, i.e. $R^*e^* = R^*e^*d^*$. Thus $Rex = R^*e^*x = R^*e^*d^*x = Redx = 0$. Therefore $eX = 0$.

Thus we have $A = H \subseteq Re$ and $e \in H$. Therefore $A = H = Re$, and by symmetry $A = G = eR$. □

Lemma 5.2. Let R be a left and right Noetherian ring which has an Artinian quotient ring Q. Set $R^* = R/N$ and let e be an idempotent element of R such that $R^*e^*$ is the socle of $R^*$. Let M be a finitely-generated torsion-free divisible right R-module then M is Artinian and M = Me.

Proof. Because M is torsion-free and divisible, i.e. for each regular element c of R we have M = Mc and the only element m of M such that mc = 0 is m = 0, we can make M in a natural way into a right Q-module. In order to prove that M is Artinian and M = Me it is enough to prove that X is Artinian and X = Xe where X is any module of the form $MN^i/MN^{i+1}$. We have NQ = QNQ = QN (1.31). Also M = MQ, so that $MN^iQ = MQN^i = MN^i$ for all i. Therefore each $MN^i$ is a right Q-module.

Set $X = MN^i/MN^{i+1}$ for some i and assume that $X \neq 0$ . We can regard X as being both a right Q-module and a right R*-module. Also X is torsion-free and divisible as a right R-module. But $\mathcal{C}(0) = \mathcal{C}(N)$ by Small's theorem (2.3(c)). Therefore X is a finitely-generated torsion-free divisible right R*-module. Let T be the quotient ring of R* then we can make X in a natural way into a right module over the semi-simple ring T. Let U be the sum of all the minimal right ideals of T which are isomorphic to T-submodules of X, then U is an ideal of T. There is a central idempotent element f of T such that U = fT. Every T-module is a direct sum of simple T-modules. Also X and U have the same isomorphism types of simple T-submodules and U is the direct sum of a finite number of simple T-modules. Therefore U is a direct summand of the direct sum of a finite number of copies of X, from which it follows that U is finitely-generated as a right R*-module. Thus there are elements $u_1,\ldots,u_n$ of U such that $U = u_1R^* + \ldots + u_nR^*$. Because T is the left quotient ring of R* there is a regular element c of R* such that $cu_j \in R^*$ for all j. Hence $cU \subseteq R^*$. But cU = cfT = fcT = fT = U. Therefore $U \subseteq R^*$. Thus an R*-submodule of X is also a U-submodule and hence also a T-submodule (because X(1 - f) = 0). But X is finitely-generated and hence Artinian as a T-module. Therefore X is Artinian as an R*-module and each simple T-submodule of X is also simple as an R*-module. Hence U is a direct sum of simple R*-modules so that $U \subseteq R^*e^*$. Hence $U = Ue^*$ so that $X = Xe^*$. Therefore X is Artinian as a right R-module and $X = Xe$. □

<u>Corollary 5.3</u>. (Robson) Let R be a left and right Noetherian ring such that $N = Nc = cN$ for all $c \in \mathcal{C}(N)$ then R is the direct sum of an Artinian ring and a semi-prime ring.

<u>Proof</u>. Because R is the direct sum of a finite number of indecomposable rings with the same properties as R, we may assume that R is indecomposable. Suppose that R is not Artinian. Let $c \in \mathcal{C}(N)$ then $N = cN$. Let $f: N \to N$ be defined by $f(n) = cn$ for all $n \in N$, then f is a surjective right R-module homomorphism. Because $r(c) \subseteq N$ we have $\mathrm{Ker}(f) = r(c)$. By Fitting's lemma there is a positive integer k such that $\mathrm{Ker}(f^k) \cap f^k(N) = 0$. Therefore $\mathrm{Ker}(f^k) = 0$ so that $\mathrm{Ker}(f) = 0$, i.e. $r(c) = 0$. Similarly $\ell(c) = 0$. Therefore $\mathcal{C}(N) \subseteq \mathcal{C}(0)$. Hence R has an Artinian quotient ring, by Small's theorem. Let e be the central idempotent given in the statement of 5.1, then, because R is indecomposable, we have either $e = 0$ or $e = 1$. But $e = 1$ if and only if R is Artinian. Therefore $e = 0$. Let c be a regular element of R then $c \in \mathcal{C}(N)$ (2.3(a)), so that $N = cN \subseteq cR$. Hence $N \subseteq eR$ (5.1(iii)). Therefore $N = 0$. □

We can now prove the decomposition theorem for left and right Noetherian p.p. rings. We shall consider p.p. rings in greater detail in Chapter 8, but for the moment all we need to know is that in a left and right Noetherian p.p. ring each right or left annihilator is generated by an idempotent. (8.4).

<u>Theorem 5.4</u>. Let R be a left and right Noetherian p.p. ring then R is a direct sum of Artinian rings and prime rings.

Proof. We begin by proving, as was first done by L.W. Small, that R has an Artinian quotient ring. Let $c \in C(N)$, then $r(c) \subseteq N$. But $r(c) = eR$ for some idempotent e. Thus $e \in N$, so that $e = 0$. Therefore $r(c) = 0$, and similarly $\ell(c) = 0$. Hence R has an Artinian quotient ring, by Small's theorem.

As in the proof of 5.3 we suppose that R is indecomposable and not Artinian. This time we must prove that R is prime. Let $x \in R$ then $\ell(x) = Re$ for some idempotent e. Hence $\ell(x)$ is an essential left ideal if and only if $e = 1$, i.e. if and only if $x = 0$. Therefore R has zero left singular ideal. Because $r(N) \neq 0$ this implies that N is not essential as a left ideal. But N is the intersection of the finitely-many minimal prime ideals of R (1.16). Therefore there is a minimal prime P of R which is not essential as a left ideal. We shall prove that $P = 0$. There is a non-zero left ideal L of R such that $P \cap L = 0$, so that $PL = 0$. Thus $P \subseteq \ell(L)$. If $xL = 0$ then $xL \subseteq P$. Also L is not contained in P. Hence $x \in P$, so that $P = \ell(L)$. Therefore $P = R(1 - f)$ for some idempotent f.

Let g be a primitive idempotent of the ring fRf, let $x \in P$, and let c be a regular element of R. We shall show that $gx \in Rc$, i.e. that $gP \subseteq Rc$. Set

$$y = f - g + gx - (1 - f)c.$$

Note that $R(1 - f)$ is an ideal of R and hence so also is fR. Thus $(1-f)Rf=0$. We also have $g = fg = gf$. Therefore $yg = fg - g^2 + gxg - (1 - f)cg = g - g + gxfg - (1 - f)cfg = 0$ because $x = x(1 - f)$. Because R has zero left singular ideal this means that Ry is not an essential left ideal. Therefore y is not left regular (1.11). Hence $\ell(y) = Rh$ for some non-zero idempotent h.

We have $hy = 0$ so that $hyf = 0$. But $hyf = hf^2 - hgf + hgxf - h(1 - f)cf = hf - hg$. Thus $hf = hg$. Set $j = hf$ then $jg = hfg = hg = hf = j$, and

$j^2 = hfhf = h^2f = hf = j$ (because $hf \in fR$). Also $j = hf = fhf$ so that $j \in fRf$. But g is a primitive idempotent of fRf. Also gj and g - gj are orthogonal idempotents in fRf. Therefore either $gj = 0$ or $gj = g$. Suppose that $gj = 0$, then $0 = jgj = j^2 = j = hf$. But $0 = hy$ so that $0 = hf - hg + hgx - h(1 - f)c = hf - hf + hfx - h(1 - f)c = -hc$. This is a contradiction because c is regular and $h \neq 0$. Therefore $gj = g$, i.e. $ghf = g$. Because $hy = 0$ we have $0 = ghy = ghf - ghg + ghgx - gh(1 - f)c = ghf - ghf + ghfx - gh(1 - f)c = gx - gh(1 - f)c$. Hence $gx \in Rc$, so that $gP \subseteq Rc$.

Let e be the central idempotent in the statement of 5.1. We are assuming that R is indecomposable and not Artinian, so that $e = 0$. But $gP \subseteq Rc$ for each regular element c of R. Therefore $gP = 0$ (5.1(iii)). But f is a sum of primitive idempotents such as g. Therefore $fP = 0$. Thus $0 = fR(1 - f) = (1 - f)Rf$, from which it follows that f is central. But R is indecomposable, so that either $f = 0$ or $f = 1$. Recall that $R(1 - f) = P$ and P is a minimal prime of R. Thus we cannot have $f = 0$. Therefore $f = 1$, i.e. $P = 0$. □

Example 5.5. Let

$$R = \begin{bmatrix} Z & Q \\ 0 & Q \end{bmatrix}$$

then R is a right Noetherian p.p. ring which is not left Noetherian (cf.4.8). It is straightforward to check that R has no non-trivial central idempotents. Thus R is indecomposable and is neither right Artinian nor prime. Also R has a right Artinian right quotient ring (namely the ring of 2 by 2 upper triangular matrices over Q), and the Artinian radical of R is a proper essential right ideal and hence is not generated by an idempotent; thus 5.1 is not true under only right-handed assumptions. Nor is 5.1 true for left and right Noetherian rings which do not have an Artinian quotient ring (4.12). □

NOETHERIAN RINGS OF KRULL DIMENSION 1.

Although we shall use the phrase "Krull dimension", no knowledge of the general theory of Krull dimension is required. Let R and S be right Noetherian rings. We say that S has the *right restricted minimum condition* if S/K is an Artinian right S-module for each essential right ideal K of S. We say that R has *right Krull dimension* 1 if R is not Artinian and R/N has the right restricted minimum condition. Thus to say that a ring has right Krull dimension 1 is, in a sense, saying that it is one step away from being right Artinian. With the usual general definition of Krull dimension it is a theorem that a right Noetherian ring R has right Krull dimension 1 if and only if R is not right Artinian and R/N has the right restricted minimum condition; for simplicity we have taken this theorem as a definition. The term "left Krull dimension 1" can be defined similarly, and we shall say that R has Krull dimension 1 if R has both right and left Krull dimension 1.

Examples. Let R be a commutative Noetherian integral domain which is not a field and which has the property that each proper factor ring of R is Artinian, then R is a Noetherian ring of Krull dimension 1. Thus any commutative Dedekind domain is a Noetherian ring of Krull dimension 1. Other examples include the ring of 2 by 2 upper triangular matrices over a commutative Noetherian domain of Krull dimension 1, the integral group ring of any finite group, and any left and right Noetherian hereditary prime ring which is not simple Artinian. □

We shall prove that the Jacobson conjecture is true for left and right Noetherian rings of Krull dimension 1, and that Noetherian rings of Krull dimension 1 with zero Artinian radical have Artinian quotient rings.

Theorem 5.6. (Lenagan) Let R be a left and right Noetherian ring of Krull dimension 1 and suppose that the Artinian radical A(R) of R is zero, then R has an Artinian quotient ring.

Proof. Let k be the smallest positive integer such that $N^k = 0$. We shall proceed by induction on k. If k = 1 then R is semi-prime and the result follows from Goldie's theorem.

Suppose now that $k \geq 2$. Set $T = N \cap \ell(N)$. The nilpotent radical of R/T is N/T. We shall show that R/T satisfies the same conditions as R and that $(N/T)^{k-1} = 0$. In fact the last statement is trivial because $N^{k-1} \subseteq \ell(N)$. Thus we must show that R/T has Krull dimension 1 and that A(R/T) = 0. Because R and hence also R/N is not Artinian, we know that R/T is not Artinian. Also (R/T)/(N/T) has the right and left restricted minimum conditions. Thus R/T has Krull dimension 1. In order to prove that A(R/T) = 0 we note that R/T can be embedded in $R/N \oplus R/\ell(N)$, so that it is enough to show that A(R/N) = 0 and $A(R/\ell(N)) = 0$. Because A(R) = 0 we have A(R/N) = 0 (4.11). Now let L be a left ideal of R such that $\ell(N) \subseteq L$ and $L/\ell(N)$ is an Artinian left ideal of $R/\ell(N)$. Let $x \in N$ then Lx is a homomorphic image of $L/\ell(N)$ because $\ell(N)x = 0$. Therefore Lx is an Artinian left ideal of R so that Lx = 0. Thus LN = 0, so that $L = \ell(N)$. Hence $A(R/\ell(N)) = 0$.

By induction on k we can suppose that R/T has an Artinian quotient ring, i.e. that the regular elements of R/T are precisely those elements of R/T which are regular modulo the nilpotent radical N/T of R/T (Small's theorem). Thus $C(N) = C(T)$. Let $c \in C(N)$. We wish to show that c is regular. Suppose that $x \in R$ and xc = 0, then $x \in T$ because $c \in C(T)$. Let f be the homomorphism from R/(cR + N) to xR defined by f(r + cR + N) = xr (f is well-defined because if $r \in cR + N$ then $xr = 0$). Thus $xR = f(R/(cR + N))$. But $c \in C(N)$ so that $(cR + N)/N$ is an essential right ideal of $R/N$. Therefore $R/(cR + N)$, and hence also $xR$, is Artinian. But $A(R) = 0$. Therefore $x = 0$, so that $\ell(c) = 0$. Similarly $r(c) = 0$. Therefore R has an Artinian quotient ring by Small's theorem (2.3(c)). □

Lemma 5.7. (Lenagan) Let R be a left and right Noetherian ring of Krull dimension 1 and suppose that $A(R) = 0$. Let K be a right ideal of R, then R/K is Artinian as a right R-module if and only if K contains a regular element.

Proof. Suppose firstly that K contains a regular element c of R. Let $a \in R$ then there exists $d \in C(N)$ such that $ad \in cR$ (2.3(b)). Thus $ad \in K$, so that R/K is torsion with respect to $C(N)$. Set $M = R/K$. We wish to show that M is Artinian, and we know that M is a finitely-generated module which is torsion with respect to $C(N)$. It is enough to show that each module of the form $MN^i/MN^{i+1}$ is Artinian.

Thus it is enough to suppose that L is a finitely-generated right R-module with $LN = 0$ and that L is torsion with respect to $C(N)$, and to prove that L is Artinian. We shall do this by showing that each cyclic submodule of L is Artinian. Let $x \in L$ then $xd = 0$ for some $d \in C(N)$. Thus $x(dR + N) = 0$, so that xR is a homomorphic image of the Artinian module $R/(dR + N)$. Therefore xR is Artinian. In this part of the proof we have not used the assumption that $A(R) = 0$.

Conversely, suppose that R/K is Artinian. We wish to show that K contains an element of $C(0)$. By 5.6, this is equivalent to showing that K contains an element of $C(N)$. Thus it is enough to show that R/K is torsion with respect to $C(N)$. But R/K has a composition series, so that it will suffice to show that each simple right R-module S is torsion. We have SN = 0 so that we can regard S as a simple right R/N-module. Thus S ≅ (R/N)/M for some maximal right ideal M of R/N. We have A(R) = 0 so that A(R/N) = 0 (4.11). Thus R/N does not contain any minimal right ideals, so that M is an essential right ideal of R/N. It now follows from Goldie's theorem that S is torsion with respect to $\mathcal{C}(N)$. □

Lemma 5.8. (Lenagan) Let R be a left and right Noetherian ring of Krull dimension 1 and let K be a right ideal of R, then R/K is Artinian if and only if K contains an element of $\mathcal{C}(A(R))$.

Proof. Suppose that R/K is Artinian. Set A = A(R). Because R/(K + A) is Artinian so also is (R/A)/((K+ A)/A). Therefore (K + A)/A contains a regular element of R/A (5.7). Thus K + A contains an element of $\mathcal{C}(A)$, from which it follows easily that K contains an element of $\mathcal{C}(A)$.

Conversely, suppose that K contains an element of $\mathcal{C}(A)$. Then (K + A)/A contains a regular element of R/A. Therefore (R/A)/((K + A)/A) is Artinian, so that R/(K + A) is Artinian. Also (K + A)/K ≅ A/(A ∩ K) and A is Artinian. Therefore R/K is Artinian. □

Corollary 5.9. (Lenagan) Let R be a left and right Noetherian ring of Krull dimension 1 and let c be a regular element of R. Then R/cR is Artinian.

Proof. This can either be proved directly as in the first part of the proof of 5.7, or it follows from 5.8 by noting that $c \in \mathcal{C}(A(R))$ because A(R) is a left and right annihilator (4.9(d)). □

Starting from the usual definition of Krull dimension (rather than the one given here) it is easy to show that if R is a ring of right Krull dimension 1 and c is a right regular element of R then R/cR is Artinian.

Before proving the Jacobson conjecture for Noetherian rings of Krull dimension 1 we shall give two examples to put the previous results into context.

Example 5.10. Set $F = Z/2Z$ and

$$R = \begin{bmatrix} Z & F \\ 0 & F \end{bmatrix} .$$

Then R is a left and right Noetherian ring of Krull dimension 1. To show that R does not have a right Artinian right quotient ring set $c = 2e_{11} + e_{22}$. Let N be the set of strictly upper triangular elements of R then N is the nilpotent radical of R. We have $c \in C(N)$ but c is not regular because $cN = 0$. Thus the condition that $A(R) = 0$ cannot be dropped from 5.6 (in fact $A(R) = N$). Let I be the set of all elements of R whose (1,1)-entry is even, then I is an ideal of R which contains no regular elements (because $IN = 0$) and R/I is finite. (cf. 5.9). □

Example 5.11. We shall give an example of a right Noetherian ring of right Krull dimension 1 which has no non-zero Artinian right ideals but which does not have a right Artinian right quotient ring. Let F be a field and let R be the ring of all matrices of the form

$$\begin{bmatrix} f(0) & g(x) \\ 0 & f(x) \end{bmatrix}$$

where x is an indeterminate and f(x) and g(x) belong to F[x]. Let t be the function which maps the matrix shown above to f(x), then t is a surjective ring homomorphism from R to F[x] and Ker(t) = N is the nilpotent radical of R. Thus $R/N \cong F[x]$, so that R/N is a Noetherian ring with the restricted minimum condition (and, incidentally, we note that N is a prime ideal of R).

Also $N = e_{12}R$ so that R is right Noetherian. We have $A(R/N) = 0$ so that any Artinian right ideal of R is contained in N. But the right R-submodules of N are determined by the ideals of F[x], and so R has no non-zero Artinian right ideals. Set $c = xe_{22}$ then $c \in C(N)$ but c is not right regular because $ce_{12} = 0$. □

THE JACOBSON CONJECTURE. When we say that a ring R satisfies the Jacobson conjecture we mean that the intersection of the powers of the Jacobson radical of R is zero. Before proving that all Noetherian rings of Krull dimension 1 satisfy the Jacobson conjecture we give a well-known example which shows that a right Noetherian ring with certain extra properties need not satisfy the Jacobson conjecture.

Example 5.12. (Herstein) Let S be Z localised at 2Z, i.e. S is the subring of Q consisting of all a/b where a and b are integers with b odd. Set $M = 2S$ then M is the Jacobson radical of S. Let

$$R = \begin{bmatrix} S & Q \\ 0 & Q \end{bmatrix} \qquad J = \begin{bmatrix} M & Q \\ 0 & 0 \end{bmatrix} .$$

Then R is a right Noetherian ring of Krull dimension 1 and J is the Jacobson radical of R. We have $J^n = M^n e_{11} + Qe_{12}$ for each positive integer n. Therefore the intersection of the powers of J is $Qe_{12}$. The fact that Q is not finitely-generated as an S-module implies that the left ideal $Qe_{12}$ of R is not finitely-generated, so that R is not left Noetherian. We note that R is a P.I. ring. □

Theorem 5.13. (Lenagan) Let R be a left and right Noetherian ring of Krull dimension 1 then R satisfies the Jacobson conjecture.

Proof. Let J and A denote the Jacobson radical and Artinian radical of R respectively. Set $J' = \bigcap_{n=1}^{\infty} J^n$. For the first part of the proof we make the extra assumption that A = 0. Let c be a regular element of R. We know that R/cR has finite length as a right R-module (5.7). Therefore there is a positive integer n such that $(R/cR)J^n = 0$, i.e. $J^n \subseteq cR$. Hence $J' \subseteq cR$ for each regular element c of R. But R has an Artinian quotient ring (5.6), and so J' = 0 (5.1(iii)).

We now turn to the general case in which we do not assume that A = 0. Because R has Krull dimension 1 and hence is not Artinian we know that R/A is not Artinian. Because the nilpotent radical of R/A contains (N + A)/A it follows easily that R/A has Krull dimension 1. Also the Artinian radical of R/A is zero (4.9). Thus the first paragraph of the proof applies to R/A so that $J' \subseteq A$. Because A is Artinian as a left R-module we know that $R/\ell(A)$ is an Artinian ring (4.5(c)). Hence $\ell(A)$ contains an element c of $\mathcal{C}(A)$ (5.8). We note that cJ' = 0 because $J' \subseteq A$. By Fitting's lemma there is a positive integer k such that $c^kR \cap r(c^k) = 0$. But $R/c^kR$ is Artinian (5.8), and as in the first paragraph of the proof we have $J' \subseteq c^kR$. Thus $J' \subseteq c^kR \cap r(c^k)$. Therefore J' = 0. □

## INJECTIVE MODULES OVER NOETHERIAN RINGS

It is easy to find examples of Artinian injective modules, e.g. every module of finite length over a semi-simple Artinian ring. It is much harder to find examples of Noetherian injective modules other than those which are Artinian. The following material on injective modules can be omitted without affecting the rest of these notes.

Theorem 5.14. Let R be a left and right Noetherian ring and let I be an injective right ideal of R, then I is Artinian.

Proof. It is enough to show that I has an essential Artinian submodule (4.7). Thus it is enough to show that each non-zero submodule K of I has a non-zero Artinian submodule. Let U be a non-zero submodule of K such that $r(U)$ is as large as possible. Set $P = r(U)$ then it is easy to show that P is a prime ideal of R. Set S = R/P then S has a simple Artinian quotient ring Q (1.28). Let $M^k$ denote the direct sum of k copies of a module M. There is a positive integer n such that S embeds as a right R-module in $U^n$ (4.5). Because Q is an essential extension of S as a right R-module, this embedding can be extended to a right R-module embedding of Q into the injective module $I^n$. But $I^n$ is finitely-generated as a right R-module. It follows that Q is finitely-generated as a right R-module and hence also as a right S-module. Therefore Q = S (1.29) so that S is an Artinian ring. Hence U is Artinian. □

The proof we have just given can easily be extended to the case where I is a finitely-generated injective right R-submodule of an R-R-bimodule which is finitely-generated as a left R-module.

Corollary 5.15. (Tachikawa) Let R be a right and left Noetherian ring which has a faithful injective right ideal I, then R is Artinian.

Proof. We know that I is Artinian (5.14) so that I is contained in the Artinian radical A of R. Because I is faithful we have $r(A) = 0$. Therefore A = R by 4.9(d). □

Remarks

(1) For the reader's convenience we give a proof (communicated to us by N.C. Norton) that *idempotents can be lifted* over a nil ideal. Let K be a nil ideal of a ring R, let $x \in R$, and assume $x - x^2 \in K$. We shall show that there is an idempotent element e of R such that $x - e \in K$. We have

$x - x^2 = k$ for some $k \in K$, and there is a positive integer n such that $k^n = 0$. We shall proceed by induction on n. Note that $kx = x^2 - x^3 = xk$. If $n > 1$, set $y = x - k + 2kx$ then $x - y \in K$ and $y - y^2 = k^2(3 + 4k)$. Thus $y - y^2 \in K$ and $(y - y^2)^{n-1} = 0$.

(2) It is not known whether a left and right Noetherian ring of right Krull dimension 1 always has left Krull dimension 1, and in particular this question is open for simple rings.

(3) 5.1 and the proof of 5.4 are taken from [25] and 5.4 first appeared in [21].

(4) Left and right Noetherian rings of Krull dimension 1 in which the Artinian radical is zero are examples of rings in which all non-zero right ideals have the same Krull dimension (such rings are now usually called K-homogeneous). For further details see, for example, [92] and [150]. For general information about Krull dimension see [67].

(5) A.V. Jategaonkar gave examples of rings in which every right ideal is principal and is a two-sided ideal and which do not satisfy the Jacobson conjecture (cf. Chapter 10).

(6) In 5.12 the intersection of the powers of the Jacobson radical of R is nilpotent, and it is an easy consequence of the two following results that this is always the case for right Noetherian P.I. rings: (Cauchon) A right Noetherian prime P.I. ring is left Noetherian; (Cauchon, Jategaonkar) A left and right Noetherian P.I. ring satisfies the Jacobson conjecture (cf. Chapter 7).

(7) Further details concerning injective Noetherian modules (and projective Artinian modules) can be found in [47].

# 6 Serial rings

Although the concept of a serial (or generalised uniserial) ring is well-known in the theory of Artinian rings, we shall follow R.B. Warfield in defining serial rings in such a way that they need not be Artinian or even Noetherian. After giving definitions and some examples, we shall prove three main results: A left and right Noetherian right serial ring satisfies the Jacobson conjecture; (Warfield) a left and right Noetherian serial ring is a direct sum of Artinian rings and prime rings; a serial ring with the a.c.c. for right annihilators has a quotient ring which is an Artinian serial ring. To prove the first two of these results we shall use the theory of the Artinian radical, and for the third we shall use the rank of a module as defined in Chapter 2 but in a non-Noetherian setting. As a bonus we shall use one of the lemmas which we prove on the way to the third main result to characterise those Noetherian rings which have a quasi-Frobenius quotient ring.

Following R.B. Warfield we define a module M to be *serial* if the set of submodules of M is linearly ordered, i.e. if, whenever A and B are submodules of M, either $A \subseteq B$ or $B \subseteq A$. A ring R is said to be *right serial* if there are orthogonal idempotents $e_1,\dots,e_n$ of R such that $R = e_1R \oplus \dots \oplus e_nR$ and each $e_iR$ is serial as a right R-module. Left serial is defined similarly, and we say that a ring is serial if it is both right and left serial.

A simple module is serial, so that any semi-simple Artinian ring is a serial ring. It is easy to show that, for example, the ring of 2 by 2 upper triangular matrices over a field is serial. We shall give some more interesting examples shortly.

Some useful facts about serial rings can be established immediately. Firstly, let I be an ideal of a right serial ring R then R/I is a right serial ring: For suppose that $R = e_1R \oplus \ldots \oplus e_nR$ where the $e_i$ are orthogonal idempotents and each $e_iR$ is serial, then $R/I \cong (e_1R/e_1I) \oplus \ldots \oplus (e_nR/e_nI)$ and each $e_iR/e_iI$ is a factor module of a serial module and so is itself serial. Secondly, let e be an idempotent element of an arbitrary ring R and assume that eR is serial as a right R-module. Let J be the Jacobson radical of R, then eJ is the intersection of all the maximal submodules of eR. Hence eJ is the unique maximal submodule of eR. Now suppose that R is right serial with $R = e_1R \oplus \ldots \oplus e_nR$ as above, then each $e_iR/e_iJ$ is a simple module from which it follows that R/J is a semi-simple Artinian ring. Thus, for each positive integer k, we know that $eJ^k/eJ^{k+1}$ is semi-simple as a right R/J-module and hence also as an R-module. Because eR is serial this implies that $eJ^k/eJ^{k+1}$ is either zero or simple. Thus $eR \supseteq eJ \supseteq eJ^2 \supseteq \ldots \supseteq eJ^k$ is a complete list of those submodules of eR which contain $eJ^k$, but in general there may be many submodules of eR which are contained in $eJ^k$ for all k.

<u>Example 6.1</u>. A left and right Artinian right serial ring need not be left serial. Let K and C be the fields of real and complex numbers respectively and set

$$R = \begin{bmatrix} K & C \\ 0 & C \end{bmatrix} .$$

Then R is a left and right Artinian ring. We have $R = e_{11}R \oplus e_{22}R$ where $e_{11}R$ has precisely one proper submodule (namely $Ce_{12}$) and $e_{22}R$ is simple. Thus R is right serial. On the other hand $Re_{22}$ is not serial as a left R-module because $Ve_{12}$ is a submodule of $Re_{22}$ for each K-subspace V of C and the K-subspaces of C are not linearly ordered. To show that R is not left serial, assume that $f_1,\dots,f_n$ are orthogonal idempotent elements of R such that $R = Rf_1 \oplus \dots \oplus Rf_n$ with each $Rf_i$ serial. We shall obtain a contradiction. Because R/N is the direct sum of two fields we must have n = 2. Since a non-zero serial module is uniform, the left Goldie dimension of R is 2. But if V and W are distinct 1-dimensional K-subspaces of C then $Ke_{11} + Ve_{12} + We_{12}$ is the direct sum of three non-zero left ideals of R, so that the left Goldie dimension of R is at least 3. Another way of proving that R is not left serial is to quote the theorem (which we shall not prove) that an indecomposable module over a left and right Artinian serial ring is serial. □

Example 6.2. Let S be Z localised at 2Z (cf. 5.12), then every proper ideal of S is of the form $2^nS$ for some positive integer n, so that S is a Noetherian serial ring which is not Artinian. The quotient ring of S is the field Q of rational numbers. We shall now show that Q is a serial S-module. (We will use this in the next example). Let a and b be relatively prime non-zero integers then either a or b is odd, i.e. either $a/b \in S$ or $b/a \in S$. Thus if x is a non-zero element of Q then either $x \in S$ or $x^{-1} \in S$, i.e. either $xS \subseteq S$ or $S \subseteq xS$. It now follows easily that if x and y are any elements of Q then either $xS \subseteq yS$ or $yS \subseteq xS$. Now let A and B be arbitrary S-submodules of Q and suppose that there is an element a of A such that $a \notin B$. Let $b \in B$. Then aS is not contained in bS, so that $bS \subseteq aS$. Therefore $B \subseteq A$. Thus Q is serial as an S-module, and in fact each proper S-submodule of Q is of the form $2^nS$ for some integer n. It follows that Q is not finitely-generated as an S-module (cf. 1.29). □

Example 6.3. Let S and Q be as in 6.2 and let

$$R = \begin{bmatrix} S & Q \\ 0 & S \end{bmatrix} .$$

Each submodule of $e_{22}R$ is of the form $Ke_{22}$ for some ideal K of S, so that $e_{22}R$ is serial. Let U be an R-submodule of $e_{11}R$ then either $U = Ae_{12}$ for some S-submodule A of Q, or $U = Be_{11} + Qe_{12}$ for some non-zero ideal B of S. Thus $e_{11}R$ is serial. Therefore R is right serial, and, by symmetry, we know that R is left serial. Although R is not Noetherian it does have the a.c.c. and d.c.c. for right annihilators, one way of justifying this being to note that R is a subring of the Artinian ring T of all 2 by 2 upper triangular matrices over Q. Also T is the quotient ring of R and T is an Artinian serial ring. □

<u>Example 6.4</u>. Let S be as in 6.2 and set

$$R = \begin{bmatrix} S & 2S \\ S & S \end{bmatrix} .$$

Then R is a left and right Noetherian ring (note that R is a finitely-generated S-module). It is straightforward to prove that R is prime. Each proper submodule of $e_{11}R$ is of the form $2^nSe_{11} + 2^nSe_{12}$ or $2^nSe_{11} + 2^{n+1}Se_{12}$ for some positive integer n, and each non-zero submodule of $e_{22}R$ is of the form $2^nSe_{21} + 2^nSe_{22}$ or $2^nSe_{21} + 2^{n+1}Se_{22}$ for some non-negative integer n. It follows easily that R is right serial, and R is left serial by symmetry. □

Recall that a complement right ideal of a ring R is a right ideal I of R such that I has no proper essential extensions in R, i.e. the only right ideal K of R which contains I as an essential submodule is K = I.

<u>Lemma 6.5</u>. Let R be any ring, let e be an idempotent element of R, and let

I be an ideal of R which is a complement as a right ideal, then eI is a complement right ideal of R.

Proof. Let K be a right ideal of R which contains eI as an essential sub-module. We have $eI \subseteq eK$ and $I + eK = (1 - e)I + eI + eK = (1 - e)I + eK$. We shall show that I + eK is an essential extension of I from which it follows that $eK \subseteq I$ and that $eI = eK$. Let x be a non-zero element of I + eK. We wish to show that $xR \cap I \neq 0$ and this is trivial if $x \in I$. Suppose that $x \notin I$ then $x = (1 - e)y + ek$ for some $y \in I$ and $k \in K$ with $ek \neq 0$. It is easily seen that eI is essential in eK. Therefore there is an essential right ideal L of R such that $ekL \neq 0$ and $ekL \subseteq eI$. Thus $xL \subseteq I$, and $xL \neq 0$ because $(1 - e)I \cap eK = 0$. Therefore $xL \neq 0$ and $xL \subseteq xR \cap I$.

At this stage we know that $eI = eK$. In particular we have $eK \subseteq K$ so that $(1 - e)K \subseteq K$. But $(1 - e)K \cap eI = 0$. Therefore $(1 - e)K = 0$, i.e. $K = eK = eI$. □

Corollary 6.6. Let R be a ring which is a direct sum of uniform right ideals and let I be an ideal of R which is a complement right ideal, then I = eR for some idempotent element e of R.

Proof. There are orthogonal idempotents $e_1, \ldots, e_n$ of R such that $R = e_1R \oplus \ldots \oplus e_nR$ and each $e_iR$ is uniform. Each $e_iI$ has no proper essential extensions in R (6.5) and $e_iR$ is uniform so that either $e_iI = 0$ or $e_iI = e_iR$. Let e be the sum of those $e_i$ for which $e_iI = e_iR$ then I = eR. □

Theorem 6.7. Let R be a left and right Noetherian right serial ring with Jacobson radical J. Then $\bigcap_{n=1}^{\infty} J^n = 0$.

Proof. Let J' denote the intersection of the powers of J. In order to prove that J' = 0 it is enough, by Nakayama's lemma, to prove that J' = J'J. By passing to the ring $R/J'J$ we may suppose that $J'J = 0$. Thus $J'$ is a finitely-generated right module over the semi-simple Artinian ring $R/J$. Hence $J'$ is contained in the Artinian radical $A$ of $R$.

Because $A$ is a complement right ideal of $R$ (4.9(c)), there is an idempotent $e$ of $R$ such that $A = eR$ (6.6). Because $eR$ has finite length as a right $R$-module we have $eR.J^k = 0$ for some positive integer $k$. Hence $eJ' = 0$. But $J' \subseteq eR$. Therefore $J' = 0$. □

Corollary 6.8. Let $R$ be a left and right Noetherian right serial ring then $R$ has the right restricted minimum condition.

Proof. Let $J$ be the Jacobson radical of $R$ and let $J'$ be the intersection of the powers of $J$, then $J' = 0$ (6.7). Let $e_1,\ldots,e_n$ be orthogonal idempotents such that $R = e_1R \oplus \ldots \oplus e_nR$ and each $e_iR$ is serial. Let $K$ be an essential right ideal of $R$. We must show that $R/K$ is an Artinian module. Because $(e_1R \cap K) \oplus \ldots \oplus (e_nR \cap K) \subseteq K$ and

$$R/((e_1R \cap K) \oplus \ldots \oplus (e_nR \cap K)) \cong e_1R/(e_1R \cap K)\oplus\ldots\oplus e_nR/(e_nR \cap K)$$

it is enough to show that each $e_iR/(e_iR \cap K)$ is Artinian. Note that $e_iR \cap K \neq 0$ because $K$ is essential.

Let $f$ be one of the $e_i$ and let $L$ be a non-zero submodule of $fR$. We must show that $fR/L$ is Artinian. Because $J' = 0$ we have $\cap fJ^k = 0$, so that there is a positive integer $k$ such that $L$ is not contained in $fJ^k$. Therefore $fJ^k \subseteq L$, so that $L = fJ^n$ for some non-negative integer $n$. (We define $J^0$ to be $R$). Thus the only submodules of $fR$ which contain $L$ are $fR$, $fJ$, $fJ^2,\ldots,$ and $fJ^n = L$, so that $fR/L$ is Artinian. □

To put 6.7 into perspective we note that Herstein's counter-example (5.12) is a right Noetherian serial ring, and that example also shows that "left

Noetherian" cannot be dropped from the statement of the next theorem.

Theorem 6.9. (Warfield) Let R be a left and right Noetherian serial ring then R is a direct sum of Artinian rings and prime rings.

Proof. We may without loss of generality assume that R is indecomposable. We further assume that R is not Artinian and we shall show that R is semi-prime and then that R is prime. Let A be the Artinian radical of R, then $A \neq R$. There are idempotents e and f of R such that A = eR = Rf (4.9 and 6.6). Therefore A is a direct summand of R, so that A = 0.

By 4.16, $\ell(N)$ is an essential right ideal. Therefore $R/\ell(N)$ is an Artinian ring (6.8) so that N is an Artinian left ideal. Therefore N = 0.

Thus, assuming that R is indecomposable and not Artinian, we have shown that R is semi-prime. Let P be a minimal prime ideal of R then P is a right annihilator (1.16). Because the right singular ideal of R is zero (1.6) it follows that P is a complement right ideal. We can now prove that P = 0 by the same method that was used in the first paragraph of the proof to show that A = 0. □

Theorem 6.10. (Chatters, Warfield) Let R be a serial ring with the a.c.c. for right annihilators then R has a quotient ring which is Artinian serial.

Proof. We are about to embark on a complicated argument. In the hope of greater clarity we will first go through the main steps of the proof in outline, and then will fill in the details in a sequence of lemmas.

The first aim is to show that the theory of the rank function defined in Chapter 2 can be applied to R even though R is not Noetherian. The nil radical N of R is nilpotent by Lanski's theorem (1.35). We know that R/N is a semi-prime serial ring and hence has finite right and left Goldie dimension, and we wish to show that R/N is a Goldie ring so we must establish the annihilator conditions. The right singular ideal Z(R) of R is nilpotent (1.6), so that $Z(R) \subseteq N$. Thus N/Z(R) is the nilpotent radical of R/Z(R). We know that R/Z(R) is a serial ring, and that R/Z(R) has the a.c.c. and d.c.c. for right annihilators (6.11). The right singular ideal of R/Z(R) is a left annihilator (6.12). We now continue the process by factoring out the right singular ideal of R/Z(R), and so on. Thus, with $Z_1 = Z(R)$, we obtain an increasing sequence $Z_1 \subseteq Z_2 \subseteq Z_3 \subseteq \ldots$ of nilpotent ideals of R such that $Z_{i+1}/Z_i$ is the right singular ideal of $R/Z_i$ and $Z_{i+1}/Z_i$ is a left annihilator in $R/Z_i$. It follows that $Z_2/Z_1 \subseteq Z_3/Z_1 \subseteq \ldots$ is an increasing sequence of left annihilators in $R/Z_1$. Therefore there is a positive integer k such that $Z_k = Z_{k+1}$. Set $T = Z_k$ and S = R/T, then S is a serial ring with zero right singular ideal and N/T is the nilpotent radical of S. Therefore R/N is a left and right Goldie ring (6.13).

We now know that N is nilpotent and that R/N is a Goldie ring. Let $\rho(M)$ denote the rank of a left R-module M. For each i let $T_i$ be the $C(N)$-torsion submodule of the left R-module $N^{i-1}/N^i$, then $T_i$ is an ideal of $R/N^i$ so that $R/T_i$ is a serial ring. Thus $R/T_i$ has finite left Goldie dimension so that $\rho(N^{i-1}/N^i)$ is finite. As in Chapter 2 we have $\rho(R) = \sum_i \rho(N^{i-1}/N^i)$ so that $\rho(R)$ is finite. Therefore a chain of left ideals of R is finite if the factors all have positive rank. We shall use this to show that R has the a.c.c. for left annihilators.

The elements of $C(N)$ are right regular (6.14). Let A and B be left annihilators in R with $A \subseteq B$ and assume that B/A has rank 0 as a left R-module. Let $b \in B$ then $cb \in A$ for some $c \in C(N)$. Thus $cb.r(A) = 0$, so that $b.r(A) = 0$. It follows that A = B. Thus if A and B are left annihilators with A strictly contained in B then B/A has positive rank. Therefore R has the a.c.c. for left annihilators. Hence the elements of $C(N)$ are also left regular (6.14).

We shall now prove the Ore condition. Let c be a regular element of R then R and cR are isomorphic as right R-modules and so have the same rank. Therefore the rank of R/cR is 0, i.e. R/cR is torsion with respect to $C(N)$. But the elements of $C(N)$ are regular. It follows that R satisfies the right Ore condition with respect to $C(0)$ and that $C(0) \subseteq C(N)$ (cf. proof of 2.3). Thus R has a right quotient ring Q which, by symmetry, is also the left quotient ring of R. To prove that Q is Artinian we note that if A and B are right ideals of Q with A strictly contained in B then $(B \cap R)/(A \cap R)$ is torsion-free and hence has positive rank as a right R-module.

To show that Q is serial, let $e_1,\dots,e_n$ be idempotents of R such that $R = e_1R \oplus \dots \oplus e_nR$ and each $e_iR$ is serial. Clearly $Q = e_1Q \oplus \dots \oplus e_nQ$ and it is enough to show that each $e_iQ$ is serial. This follows easily because $e_iR$ is serial and if A is a right ideal of Q then $A = (A \cap R)Q$. □

We shall now fill in the gaps in the proof of Theorem 6.10.

<u>Lemma 6.11</u>. (Fisher) Let R be any ring, let Z denote the right singular ideal of R, and assume that R/Z has a chain of n + 1 distinct right annihilators for some positive integer n, then R contains a direct sum of n non-zero right ideals.

<u>Proof</u>. By assumption, there are right ideals $B_1,\dots,B_{n+1}$ of R such that $Z \subseteq B_1 \subsetneqq B_2 \subsetneqq \dots \subsetneqq B_{n+1}$ and each $B_i/Z$ is a right annihilator in R/Z. For each i let $A_i$ be the left ideal of R which contains Z such that $A_i/Z$ is the left annihilator in R/Z of $B_i/Z$. Thus $Z \subseteq A_{n+1} \subsetneqq A_n \subsetneqq \dots \subsetneqq A_1$ and $A_i$ is the largest left ideal such that $A_iB_i \subseteq Z$. For each i from 1 to n fix $a_i \in A_i$ and $b_{i+1} \in B_{i+1}$ such that $a_ib_{i+1} \notin Z$.

For each i from 2 to n + 1 we define a statement P(i) as follows:

P(i) = "There exist right ideals $K_2,\dots,K_i$ such that whenever $2 \le j \le i$, we have

(a) $K_j \cap r(b_j) = 0$,

(b) $b_jK_j \cap r(a_{j-1}) = 0$, and

(c) $r(a_i) \cap r(a_{i-1}) \cap \dots \cap r(a_j) \cap b_jK_j \neq 0$."

By induction we shall prove that P(i) is true for all i.

To show that P(2) is true: Because $a_1b_2 \notin Z$ we know that $r(a_1b_2)$ is not an essential right ideal. Hence there is a non-zero right ideal $K_2$ with $r(a_1b_2) \cap K_2 = 0$. We have $r(b_2) \cap K_2 \subseteq r(a_1b_2) \cap K_2$ so that (a) is satisfied. Let $k_2 \in K_2$ with $b_2k_2 \in r(a_1)$ then $k_2 \in r(a_1b_2) \cap K_2$ so that $k_2 = 0$ and (b) is satisfied. To verify (c) we must show that $r(a_2) \cap b_2K_2 \neq 0$. Because $r(a_2b_2)$ is an essential right ideal we have $r(a_2b_2) \cap K_2 \neq 0$, i.e. $a_2b_2x = 0$ for some non-zero element x of $K_2$. But $b_2x \neq 0$ by (a), so that $b_2x$ is a non-zero element of $r(a_2) \cap b_2K_2$. From now on we shall omit this kind of verification.

Now suppose that P(i) is true for some i from 2 to n. We shall show that P(i + 1) is true. Let $K_2,\dots,K_i$ be the right ideals given by P(i). Because $r(a_ib_{i+1})$ is not essential, there is a non-zero right ideal $K_{i+1}$ such that $r(a_ib_{i+1}) \cap K_{i+1} = 0$. With this choice of $K_{i+1}$ it is easy to show that (a) and (b) are true for P(i + 1). To deal with (c) we first note that $K_{i+1} \cap r(a_{i+1}b_{i+1}) \neq 0$ so that $r(a_{i+1}) \cap b_{i+1}K_{i+1} \neq 0$ (using (a)). Thus P(i + 1)(c) is true for j = i + 1. Now fix j with $2 \le j \le i$. Because P(i) is true there is an element $k_j$ of $K_j$ such that $b_jk_j \neq 0$ and $b_jk_j \in r(a_i) \cap \dots \cap r(a_j)$. Because $a_{i+1}b_j \in Z$, there is an element x of R such that $k_jx \neq 0$ and $k_jx \in r(a_{i+1}b_j)$. Thus $b_jk_jx \neq 0$ (by (a)) and $b_jk_jx \in r(a_{i+1}) \cap r(a_i) \cap \dots \cap r(a_j)$, so that (c) is true.

Therefore P(i) is true for all i from 2 to n + 1. It follows easily that each of the right ideals $b_{n+1}K_{n+1}$, $b_nK_n \cap r(a_n)$, $b_{n-1}K_{n-1} \cap r(a_n) \cap r(a_{n-1})$, ..., $b_2K_2 \cap r(a_n) \cap r(a_{n-1}) \cap \ldots \cap r(a_2)$ is non-zero (by P(n + 1)(c) and P(n)(c)) and their sum is direct (by P(n + 1)(b)). □

Lemma 6.12. Let R be a ring with the d.c.c. for right annihilators then the right singular ideal Z of R is a left annihilator.

Proof. Let $z_1 \in Z$ and set $A_1 = r(z_1)$. We have $\ell(A_1) \subseteq Z$ because $A_1$ is an essential right ideal. If $\ell(A_1) = Z$ we stop. If not, there exists $z_2 \in Z$ such that $z_2A_1 \neq 0$. Set $A_2 = r(z_1,z_2) = r(z_1) \cap r(z_2)$. Then $A_2$ is an essential right ideal and is a right annihilator. Also $A_2 \subsetneq A_1$ because $z_2A_2 = 0$ and $z_2A_1 \neq 0$. If $\ell(A_2) = Z$ we stop. If not, there exists $z_3 \in Z$ such that $z_3A_2 \neq 0$. Set $A_3 = r(z_1,z_2,z_3)$, and so on. □

Lemma 6.13. Let R be a right non-singular right serial ring then R/N is a right Goldie ring.

Proof. We know that R has the a.c.c. and d.c.c. for right annihilators (1.14). Let P be an ideal of R which is maximal with respect to being of the form $\ell(I)$ for some non-zero ideal I of R, then P is prime (P is called a *maximal left annihilator* prime). Because R is right non-singular we know that $r(P)$ is a complement right ideal of R so that $r(P) = eR$ for some idempotent e (6.6). Set S = eRe, f = 1 - e, and T = fRf. Both eR and Rf are ideals of R so that Re = eRe and fR = fRf. The function given by $x \to fxf$ is a surjective ring homomorphism from R to T with kernel eR so that $T \cong R/eR$. Thus T is a right serial ring. Similarly S is right serial. In fact $S \cong R/P$ so that S is a prime ring. To show that T is right non-singular, let K be an essential right ideal of T then it is easy to show that eR + K is an essential right ideal of R. If $fxfK = 0$ then $fxf(eR + K) = 0$ so that $fxf = 0$.

Thus S is a prime right serial ring and, being a subring of R has the a.c.c. for right annihilators. Also T is a right non-singular right serial ring. The nilpotent radical of T is fNf where N is the nilpotent radical of R. Right ideals of T are right ideals of R, and T is not essential when regarded as a right ideal of R. Therefore the right Goldie dimension of T is strictly less than that of R, so that by induction we can assume that T/fNf is a right Goldie ring.

Since both S and T/fNf are right Goldie rings, to complete the proof it is enough to show that $R/N \cong S \oplus T/fNf$. Let $g:R \to (S \oplus T/fNf)$ be defined by $g(x) = (exe, fxf + fNf)$ then g is a surjective ring homomorphism. Let $K = Ker(g)$, then clearly $N \subseteq K$. Because $g(K) = 0$ we have $eKe = 0$ and $fKf \subseteq fNf$, i.e. $Ke = 0$ and $fK \subseteq N$. Thus eK is nilpotent, so that $eK \subseteq N$ and $fK \subseteq N$. Therefore $K \subseteq N$, so that $K = N$. □

Lemma 6.14. (Hajarnavis and Ludgate) Suppose that both R and R/N are right Goldie rings and that R contains a direct sum of uniform right ideals which contains a right regular element, then the elements of $C(N)$ are right regular.

Proof. There are uniform right ideals $U_1,\dots,U_n$ of R such that the sum $U_1 + \dots + U_n$ is direct, and $u_1 + \dots + u_n$ is right regular for some $u_i \in U_i$. Thus $r(u_1) \cap \dots \cap r(u_n) = 0$. Let K be a right ideal of R which has non-nilpotent intersection with each non-nilpotent right ideal of R. We shall show that K contains a right regular element, and eventually we will take $K = cR + N$ with $c \in C(N)$.

Set $u = u_1 + \dots + u_n$ and suppose that $u_1 \in N$, then $u - u_1$ is right regular (6.15). Thus the right ideal $U_2 \oplus \dots \oplus U_n$ contains a right regular element and so is essential (1.11), which is a contradiction. Similarly for each i we have $u_i \notin N$. Thus each $u_iR$ is non-nilpotent. Set $V_i = K \cap u_iR$ then $V_i$ is not nilpotent. Thus $V_iu_i$ is not contained in N, so that $v_iu_i \notin N$ for some $v_i \in V_i$. Hence $r(v_iu_i)$ is not an essential right ideal of R (1.6). Clearly $r(u_i) \subseteq r(v_iu_i)$. We shall prove that $r(u_i) = r(v_iu_i)$ by showing that any right ideal which strictly contains $r(u_i)$ is essential. Let A be such a right ideal and let B be a right ideal with $A \cap B = 0$. We have $u_iA \neq 0$. Let $x \in u_iA \cap u_iB$ then $x = u_ia = u_ib$ for some $a \in A$ and $b \in B$. We have $u_i(a - b) = 0$ so that $a - b \in r(u_i)$. Hence $a - b \in A$ so that $b \in A$ and hence $b = 0$. Therefore $u_iA \cap u_iB = 0$, and, because $u_iR$ is uniform, this gives $u_iB = 0$. It follows that $B \subseteq A$ so that $B = 0$.

Thus $r(u_i) = r(v_iu_i)$ for all i. Set $k = v_1u_1 + \ldots + v_nu_n$ then $k \in K$, and k is right regular because the sum $v_1u_1R + \ldots + v_nu_nR$ is direct and $r(v_1u_1) \cap \ldots \cap r(v_nu_n) = r(u_1) \cap \ldots \cap r(u_n) = 0$.

Now let $c \in C(N)$ and set $K = cR + N$. Because K/N is an essential right ideal of R/N we know that K has non-nilpotent intersection with each non-nilpotent right ideal of R. Therefore K contains a right regular element, and hence so also does cR (6.15). Thus cd is right regular for some $d \in R$. Thus $r(c) = 0$ (1.30(b)). □

<u>Lemma 6.15</u>. (Goldie) Let R be a ring with the a.c.c. for right annihilators and assume that xR is an essential right ideal of R whenever x is a right regular element of R. Let N be a nilpotent ideal of R, let $n \in N$, and let c be a right regular element of R, then $c - n$ is right regular.

<u>Proof</u>. Let i be a fixed positive integer. We have $r(N^ic) \subseteq r(N^ic^2) \subseteq \ldots$ so that there is a positive integer k (depending on i) such that $r(N^ic^k) = r(N^ic^j)$ for all $j \geq k$. Because there are only finitely many values of i for which $N^i \neq 0$, it follows that there is a positive integer k such that

$r(N^i c^k) = r(N^i c^j)$ for all $j \geq k$ and all i. Set $d = c^k$, then $r(N^i d) = r(N^i d^2)$ for all i. In order to prove that c - n is right regular it is enough to show that $r(c^{k-1}(c - n)) = 0$, i.e. that $r(d - c^{k-1}n) = 0$. Clearly d is right regular and $c^{k-1}n \in N$.

Thus without loss of generality we may suppose that $r(N^i c) = r(N^i c^2)$ for all i. Let $x \in R$ and suppose that $cx \in r(c - n)$. Thus $c^2 x = ncx$. Let t be a positive integer such that $N^t = 0$ then $0 = N^{t-1}ncx = N^{t-1}c^2 x$. Thus $x \in r(N^{t-1}c^2)$ so that $N^{t-1}cx = 0$ because $r(N^{t-1}c^2) = r(N^{t-1}c)$. Hence $N^{t-2}ncx = 0$ so that $N^{t-2}c^2 x = 0$, and so on. Eventually we obtain $Nc^2 x = 0$ so that $Ncx = 0$. Hence $0 = ncx = c^2 x$ so that $cx = 0$.

Therefore $cR \cap r(c - n) = 0$. But c is right regular so that cR is essential. Hence $r(c - n) = 0$. □

We have now finished everything to do with the proof of 6.10. From the assumptions that R is right serial with the a.c.c. for right annihilators we proved that elements of $C(N)$ are right regular. The assumption that R is also left serial was used to show that the elements of $C(N)$ are also left regular. The following example shows that right-handed assumptions alone are not enough to give an Artinian quotient ring in these circumstances.

Example 6.16. The following is a left and right Noetherian right serial ring which is its own quotient ring but which is not Artinian. Let S be Z localised at 2Z and let M = 2S then $S/M \cong Z/2Z$. Set F = S/M then F is, in a natural way, an S-module. Let

$$R = \begin{bmatrix} F & F \\ 0 & S \end{bmatrix} .$$

Then R is left and right Noetherian right serial. Because $Fe_{12}$ and $Me_{22}$ are incomparable left R-submodules of $Re_{22}$ it follows that R is not left serial (cf. 6.1). Because $FM = 0$, an element of R is regular if and only if its (2,2)-entry is an element of S which does not belong to M. Since such elements of S are units of S it follows that the regular elements of R are units of R. Thus R is its own quotient ring, and R is not Artinian because S is a homomorphic image of R. □

As another application of 6.14 we shall give a characterisation of those left and right Noetherian rings which have a quotient ring which is a QF (i.e. quasi-Frobenius) ring. We shall not go into details about QF rings except to say that they can be defined as those Artinian rings which are self-injective, or as those Artinian rings in which every right ideal is a right annihilator and every left ideal is a left annihilator. The group algebra of any finite group over any field is a QF ring, and so also is each proper factor ring of a Dedekind domain.

Theorem 6.17. (Kupisch) Let R be a left and right Artinian ring then R is a QF ring if and only if

(i) R is a direct sum of uniform right ideals,

(ii) R is a direct sum of uniform left ideals, and

(iii) the right and left socles of R are equal.

Theorem 6.18. Let R be a left and right Noetherian ring then R has a QF quotient ring if and only if

(i) R contains a direct sum of uniform right ideals which contains a right regular element,

(ii) R contains a direct sum of uniform left ideals which contains a left regular element, and

(iii) $\ell(N) = r(N)$.

Proof. Suppose that R satisfies (i), (ii) and (iii), then $C(N) \subseteq C(0)$ (6.14), so that R has an Artinian quotient ring Q by Small's theorem (2.3(c)). It is straightforward to show that conditions (i) and (ii) also hold in Q. But right or left regular elements of Q are units of Q. Therefore Q is a direct sum of uniform right ideals and of uniform left ideals. Let N' be the nilpotent radical of Q, then $N' \cap R$ is a nilpotent ideal of R so that $N' \cap R \subseteq N$. Thus $N' \subseteq NQ$. But NQ is an ideal of Q (1.31), from which it follows easily that NQ is nilpotent. Hence $NQ \subseteq N'$ so that $N' = NQ$. By symmetry we have $N' = QN$. It follows easily from (iii) that the left and right annihilators of N' in Q are equal, and because Q is Artinian this means that the right and left socles of Q are equal. Therefore Q is a QF ring (6.17).

We leave the converse as an exercise. □

Remarks

(1) For further details about serial rings, including their module theory, see [155].

(2) For further details concerning Artinian serial rings see, for example, [40].

(3) In 8.15 we shall prove that a right non-singular right serial ring is right semi-hereditary.

(4) Theorem 6.10 appears in [156].

(5) Lemma 6.15 appears in [58].

(6) Let R be a serial ring with the a.c.c. for right annihilators. We proved in 6.10 that R has a quotient ring Q which is Artinian serial. We shall now show that R and Q have the same nilpotent radical. We did not include this in the main part of the text because we do not know a proof which is self-contained in the context of these notes. Let N and N' be the

nilpotent radicals of R and Q respectively and let $c \in C(N)$. By Theorem 3.3 of [155], there are orthogonal idempotents $e_1,\dots,e_n$ of R such that $R = e_1R \oplus \dots \oplus e_nR$, each $e_iR$ is serial, and $cR = (e_1R \cap cR) \oplus \dots \oplus (e_nR \cap cR)$. For each i we have either $e_iN \subseteq e_iR \cap cR$ or $e_iR \cap cR \subseteq e_iN$. But $(cR + N)/N$ is an essential right ideal of R/N (1.11) so that $e_iR \cap (cR + N)$ is not nilpotent. But $e_iR\cap(cR+N) = e_iR\cap(e_1R\cap cR + e_1N+\dots+e_nR\cap cR+e_nN)=e_iR\cap cR + e_iN$. Therefore $e_iR \cap cR \not\subseteq e_iN$, so that $e_iN \subseteq e_iR \cap cR$ for all i. Hence $N \subseteq cR$, so that $N = N \cap cR = cN$. But c is a unit of Q so that $N = c^{-1}N$. Since every element of Q is of the form $ac^{-1}$ for some $a \in R$ and $c \in C(N)$, we have $N' = QN = N$. This proof that $N = cN$ can be used as an alternative way of proving that if $c \in C(N)$ then $r(c) = 0$ (cf. proof of 5.3).

(7) For further details concerning QF rings see, for example, [41]. However, it is interesting to note that a left and right Artinian ring in which every right ideal is a right annihilator need not be a QF ring [34]. Also the assertion that every proper factor ring of a Dedekind domain is a QF ring is true for non-commutative Dedekind rings [117].

(8) An obvious consequence of 6.11 is that if R is a ring with finite right Goldie dimension and Z denotes the right singular ideal of R then R/Z has the a.c.c. and d.c.c. for right annihilators. We shall now sketch an alternative proof of this due to R. Gordon and J.C. Robson. Let E denote the injective hull of R as a right R-module then E has finite Goldie dimension. Let S be the endomorphism ring of E, then the Jacobson radical J of S consists of those elements of S whose kernels are essential submodules of E. We can consider R to be a subring of S because left multiplication by an element of R gives rise to an endomorphism of E. We have $Z = R \cap J$, so that R/Z embeds in the semi-simple Artinian ring S/J.

(9) It is not known whether a semi-prime right serial ring is always a right Goldie ring.

# 7 Fully bounded rings

We shall give two little-known proofs of the Jacobson conjecture for Noetherian fully bounded rings. Both proofs use the material of Chapter 4, and one of them uses the "Moss prime". We shall also give a proof due to A.W. Goldie of G. Cauchon's result that a right Noetherian right fully bounded ring satisfies the H-condition. The chapter ends with an unpublished example due to I.M. Musson of a left and right Noetherian ring R which has a simple module V and a finitely-generated essential extension M of V such that M is not Artinian (7.15).

Let R be any ring. We say that R is *right bounded* if every essential right ideal of R contains a two-sided ideal of R which is essential as a right ideal. Thus if R is prime then R is right bounded if and only if each essential right ideal of R contains a non-zero ideal of R. A ring R is said to be *right fully bounded* if each prime factor ring of R is right bounded.

A simple Artinian ring has no proper essential right ideals and so is right bounded, and it follows that any right Artinian ring is right fully bounded. In fact a simple ring is right bounded if and only if it is Artinian, so that all simple factor rings of a right fully bounded ring are Artinian. It was shown by S.A. Amitsur that a prime ring which satisfies a polynomial identity is bounded.

Recall that if S is a non-empty subset of a right R-module M we write

$$r(S) = \{x \in R : sx = 0 \text{ for all } s \in S\}.$$

We call M *faithful* if $r(M) = 0$. The singular submodule of M is

$Z(M) = \{m \in M: r(m)$ is an essential right ideal of $R\}$.

We say that M is non-singular if $Z(M) = 0$.

Lemma 7.1. Let R be a prime right Noetherian right bounded ring and let U be a finitely-generated uniform faithful right R-module then $Z(U) = 0$.

Proof. Suppose that $Z(U) \neq 0$. Then $Z(U)$ is an essential submodule of U. Let $x_1,\ldots,x_n$ be a finite generating set for U. For each i there is an essential right ideal $L_i$ with $x_iL_i \subseteq Z(U)$ (1.1). Set $L = L_1 \cap \ldots \cap L_n$ then, by Goldie's theorem, we know that L contains a regular element c. Thus $x_ic \in Z(U)$ for all i. Similarly there is a regular element d of R such that $x_icd = 0$ for all i. But cdR contains a non-zero ideal I of R. Therefore $x_iI = 0$ for all i so that $UI = 0$, which is a contradiction. □

The following example shows that, in general, a finitely-generated faithful uniform module can have non-zero singular submodule.

Example 7.2. Let R be a simple Noetherian ring which is not Artinian, e.g. let R be the ring of polynomials over the complex numbers in indeterminates x and y subject to the relation $xy - yx = 1$ (this is the first Weyl algebra). Let M be a maximal right ideal of R. Then R/M is a cyclic faithful uniform (indeed simple) right R-module. Because R has no minimal right ideals (1.24) we know that M is an essential right ideal of R. Therefore $Z(R/M) = R/M$ (1.1). □

Lemma 7.3. (Ginn and Moss) Let R be a left and right Noetherian ring and let M be a right R-module such that $R/r(M)$ is not Artinian. Let P be an ideal of R which is maximal with respect to the properties that R/P is not Artinian and $P = r(L)$ for some submodule L of M. Then P is a prime ideal. We call P a *Moss prime* associated with M.

*Proof*. Let A and B be ideals of R both of which strictly contain P and suppose that $AB \subseteq P$. We have $P = r(L)$ for some submodule L of M. Because $LAB = 0$ we can take $B = r(LA)$. By maximality of P we know that R/B is Artinian. Therefore there is an ideal C of R such that R/C is Artinian and $CA \subseteq AB$ (4.4). Thus $LCA = 0$. Set $D = r(LC)$ then $A \subseteq D$. Hence R/D is Artinian. We have $LCD = 0$ so that $CD \subseteq P$. But R/C is Artinian, and C/CD is a finitely-generated module over the Artinian ring R/D. It follows that R/CD and hence also R/P is Artinian, which is a contradiction. □

*Lemma 7.4*. Let R be a ring with the property that each cyclic uniform right R-module with a simple submodule has finite length. Let J be the Jacobson radical of R. Then $\bigcap_{n=1}^{\infty} J^n = 0$.

*Proof*. Set $J' = \bigcap_{n=1}^{\infty} J^n$. Let x be a non-zero element of R. By Zorn's lemma there is a right ideal $I_x$ of R such that $I_x$ is maximal with respect to not containing x. Each non-zero submodule of $R/I_x$ contains $x + I_x$. Hence $R/I_x$ is uniform and $(xR + I_x)/I_x$ is a simple submodule. Therefore $R/I_x$ has finite length, so that $(R/I_x)J^n = 0$ for some n. Thus $J^n \subseteq I_x$ so that $J' \subseteq I_x$. But $\cap I_x = 0$ as x ranges over the non-zero elements of R. Therefore $J' = 0$. □

*Theorem 7.5*. (Cauchon, Jategaonkar) Let R be a left and right Noetherian right fully bounded ring with Jacobson radical J. Then $\bigcap_{n=1}^{\infty} J^n = 0$.

*Proof*. The following proof is due to S.M. Ginn and P.B. Moss.
Let M be a cyclic uniform right R-module with a simple submodule S and assume that M is not Artinian. Because M is a cyclic $R/r(M)$-module we know that $R/r(M)$ is not Artinian. Let P be a Moss prime associated with M (7.3) then $P = r(L)$ for some submodule L of M. Clearly $L \neq 0$ so that $S \subseteq L$. Set $T = R/P$ then L is a finitely-generated faithful uniform right T-module.

Considering S as a right T-module we have Z(S) = 0 because Z(L) = 0 (7.1). But S ≅ T/K for some maximal right ideal K of T, so that K is not an essential right ideal of T (1.1). It follows that T has a minimal right ideal and so is Artinian (1.24). This is a contradiction, so that M is Artinian and we can use 7.4. □

The Jacobson conjecture is not true for right Noetherian right fully bounded rings (5.12).

We shall now derive a module-theoretic characterisation of right Noetherian right fully-bounded rings which in particular will be used to give another proof of 7.5.

A ring R is said to satisfy the H-*condition* if for each finitely-generated right R-module M there is a finite subset S of M such that $r(S) = r(M)$, or equivalently there is a positive integer n such that $R/r(M)$ embeds in $M^n$. It is easy to show that if R has the H-condition then so also does every factor ring of R.

Proposition 7.6. Let R be a ring with the H-condition then every factor ring of R is right bounded.

Proof. It is enough to show that R is right bounded. Let K be an essential right ideal of R and set M = R/K. We have $r(M) = r(S)$ for some finite subset S of M. Hence there are finitely-many elements $y_1,\ldots,y_n$ of R such that if $x \in R$ then $Rx \subseteq K$ if and only if $y_i x \in K$ for all i. Let $L_i = \{x \in R: y_i x \in K\}$. Then $L_i$ is an essential right ideal of R (1.1). Set $L = L_1 \cap \ldots \cap L_n$. Then $L = r(M)$. Clearly $r(M)$ is an ideal of R and $r(M) \subseteq K$. Also L is an essential right ideal because each $L_i$ is essential. □

The next aim is to prove the converse of 7.6 for right Noetherian rings (7.8).

Let R be a right Noetherian ring and let U be a uniform right R-module. Let P be an ideal of R which is maximal with respect to being of the form $P = r(V)$ for some non-zero submodule V of U. If X and Y are non-zero submodules of U then $X \cap Y \neq 0$ and $r(X) + r(Y) \subseteq r(X \cap Y)$. It follows that $r(X) \subseteq P$ for each non-zero submodule X of U and that P is a prime ideal of R. We call P the *assassinator* of U and write $P = \text{ass}(U)$.

Lemma 7.7. Let R be a right Noetherian right fully bounded ring and let U be a uniform faithful right R-module. Set $P = \text{ass}(U)$. Then $\ell(P)$ is an essential right ideal of R.

Proof. It is enough to show that $\ell(P)$ has non-zero intersection with each uniform right ideal of R. Let X be a uniform right ideal of R and set $P' = \text{ass}(X)$. There is a non-zero submodule Y of X such that $r(Y) = P'$. Because U is faithful we have $UY \neq 0$ and $r(UY) = r(Y) = P'$. Hence $P' \subseteq P$. We have $P = r(V)$ for some non-zero submodule V of U. Set $W = V \cap UY$. We know that UY is non-singular as a right $R/P'$-module (7.1). But W is a non-zero submodule of UY and $WP = 0$. Therefore $P/P'$ is not an essential right ideal of $R/P'$ and it follows that $P = P'$. Hence $YP = 0$ so that $\ell(P) \cap X \neq 0$. □

Theorem 7.8. (Cauchon) Let R be a right Noetherian right fully bounded ring. Then R has the H-condition.

Proof. The following proof is due to A.W. Goldie. Let M be a finitely-generated right R-module. There are submodules $K_1, \ldots, K_n$ of M such that $K_1 \cap \ldots \cap K_n = 0$ and each $M/K_i$ is uniform (7.9). We have $r(M) = r(M/K_1) \cap \ldots \cap r(M/K_n)$. In order to show that $r(M) = r(S)$ for some finite subset S of M it is enough to show that for each i there is a finite subset $S_i$ of $M/K_i$ such that $r(M/K_i) = r(S_i)$ (choose S to be a finite subset of M such that the image of S in $M/K_i$ contains $S_i$).

Thus without loss of generality we may suppose that M is uniform. By passing to the ring $R/r(M)$ we may also suppose that M is faithful. Let $P = \text{ass}(M)$ then $\ell(P)$ is an essential right ideal of R (7.7). Set $L = \{m \in M: mP = 0\}$ then $r(L) = P$. Thus L is non-singular and hence torsion-free as a right R/P-module (7.1). Let $\rho(X)$ denote the rank of a right R/P-module X. Then $\rho(L') = 1$ for each non-zero submodule L' of L because L is torsion-free and uniform.

Let $x_1 \in M$ and set $A_1 = \ell(P) \cap r(x_1)$. If $A_1 = 0$ then $r(x_1) = 0$, so that $r(x_1) = r(M)$ and we can stop. Suppose now that $A_1 \neq 0$ then there exists $x_2 \in M$ such that $x_2A_1 \neq 0$. We have $x_2A_1P = 0$ so that $x_2A_1 \subseteq L$. Thus $\rho(x_2A_1) = 1$. Set $A_2 = A_1 \cap r(x_2) = \ell(P) \cap r(x_1,x_2)$. Let $f:A_1 \to x_2A_1$ be given by $f(a) = x_2a$ for all $a \in A_1$. Then $\text{Ker}(f) = A_2$. Thus $A_1/A_2 \cong x_2A_1$ so that $\rho(A_1/A_2) = 1$. If $A_2 \neq 0$ there exists $x_3 \in M$ such that $x_3A_2 \neq 0$. Set $A_3 = A_2 \cap r(x_3)$ then as above we have $\rho(A_2/A_3) = 1$. This process produces a chain of submodules of $A_1$ such that the corresponding factor modules have rank 1, and hence it must stop after at most $\rho(A_1)$ steps. Therefore there are elements $x_1,\ldots,x_n$ of M such that $\ell(P) \cap r(x_1,\ldots,x_n) = 0$. Hence $r(x_1,\ldots,x_n) = 0 = r(M)$. □

For the purposes of the next result only we define a submodule K of a module M to be *co-uniform* if M/K is uniform.

Lemma 7.9. Let M be a Noetherian module. Then every submodule of M is the intersection of a finite number of co-uniform submodules of M.

Proof. Suppose the result is false and let A be a submodule of M which is maximal with respect to not being the intersection of a finite number of co-uniform submodules of M. Because M/A is not uniform there are submodules B and C of M which strictly contain A such that $A = B \cap C$. By maximality of

A we know that each of B and C is a finite intersection of co-uniform submodules of M and hence so also is A. This is a contradiction. □

Theorem 7.10. (Jategaonkar) Let R be a left and right Noetherian right fully bounded ring and let M be a finitely-generated right R-module which has an essential Artinian submodule. Then M is Artinian.

Proof. There is a positive integer n such that $R/r(M)$ embeds in $M^n$ (7.8 and 4.5(a)). But M has an essential Artinian submodule and hence so also do $M^n$ and $R/r(M)$. Therefore $R/r(M)$ is Artinian (4.6). □

Theorem 7.11. (Cauchon, Jategaonkar) Let J be the Jacobson radical of a left and right Noetherian right fully bounded ring. Then $\bigcap_{n=1}^{\infty} J^n = 0$.

Proof. Combine 7.10 and 7.4. □

Proposition 7.12. Let R be a right Noetherian right fully bounded ring then R is right bounded.

Proof. Combine 7.8 and 7.6. □

Proposition 7.13. Let R be a left and right Noetherian right fully bounded ring and assume that there is a non-zero ideal I of R such that every non-zero ideal of R contains I, then R is Artinian.

Proof. As in the proof of 7.4, there is a family of right ideals K of R whose intersection is zero such that each R/K is a uniform module with a simple submodule. At least one K does not contain I and hence R/K is faithful. For such a K we know that R/K is Artinian (7.10) and R embeds in $(R/K)^n$ for some positive integer n (7.8 and 4.5(a)). Therefore R is Artinian. □

The assumption that R is right fully bounded cannot be dropped from 7.13

(4.18).

We shall now give two examples concerning essential extensions of Artinian modules (cf. 7.10). It was shown by A.V. Jategaonkar that if R is a left and right Noetherian P.I. ring and M is a (not necessarily finitely-generated) essential extension of a simple module then M is Artinian. The following example shows that this result cannot be extended to Noetherian fully bounded rings.

Example 7.14. It was shown by P.M. Cohn that there is a division ring D with a subdivision ring K such that D is finite-dimensional as a left vector space over K but not as a right vector space over K. Let

$$R = \begin{bmatrix} K & D \\ 0 & D \end{bmatrix} .$$

Then the nilpotent radical of R is $N = De_{12}$. We have $N = e_{12}R$, and N is finitely-generated as a left ideal of R because D is a finitely-generated left K-module. It follows easily that R is left and right Artinian. Set $M = De_{11} + De_{12}$ then M is a right R-module which has the simple module N as an essential submodule. The R-submodules of M/N correspond to the right K-subspaces of D, so that M is not Artinian. □

We are very grateful to I.M. Musson for allowing us to include the following unpublished example of a left and right Noetherian ring which has a non-Artinian cyclic (in fact serial) essential extension of a simple module.

Example 7.15. (Musson) Let C be the field of complex numbers and let R be the ring of polynomials over C in indeterminates x and y subject to the relation $xy - yx = x$. In fact R is the universal enveloping algebra of a certain finite dimensional Lie algebra and so, by standard results, is a left and right Noetherian integral domain. The elements of R can be expressed uniquely as finite sums of the form $\Sigma a_{ij}x^iy^j$ where i and j take non-negative integer values and $a_{ij} \in C$. It follows easily that $xR = Rx$. Set $I = (y - 1)(x - 1)R$ and $M = R/I$. For each non-negative integer n set $v_n = (y - 1)y^n + I$ then $v_n = v_oy^n$. We have $v_ox = v_o$ because $(y - 1)x - (y - 1) \in I$. Set $V = v_oR$ then $v_n \in V$ for all n.

The first aim is to show that V is simple as a right R-module. We shall show by induction that

(i) $v_n(1 - x)^n = v_on!$.

Suppose that $v_n(1-x)^n = v_on!$. Then $v_{n+1}(1 - x)^{n+1} = v_ny(1 - x)^{n+1} = v_n(y - yx)(1 - x)^n = v_n(y - xy + x)(1 - x)^n = v_n(1 - x)y(1 - x)^n + v_nx(1 - x)^n = v_n(1 - x)(y - xy + x)(1 - x)^{n-1} + v_on! = v_n(1 - x)^2y(1 - x)^{n-1}+2v_on!=\ldots= v_n(1-x)^{n+1}y + (n + 1)v_on! = v_o(1 - x)yn! + v_o(n + 1)! = v_o(n + 1)!$. Thus (i) is true for all n. Because $v_ox = v_o$ each element of $v_oR$ is of the form $v = v_oa_o + \ldots + v_na_n$ with $a_i \in C$. Let v be as above with $a_n \neq 0$ then $v(1 - x)^n = v_on!a_n$ so that $v_o \in vR$. Therefore V is simple. Another easy consequence of (i) is that the $v_i$ form a basis for V as a vector space over C.

Now let $w_n = x^n + I$ for each non-negative integer n then $M = w_oR$ and $w_n = w_ox^n$. It is trivial to show by induction that $x^ny = yx^n + nx^n$ so that $w_ny = w_ox^ny = w_oyx^n + nw_ox^n$. But $w_oy = (1 + I)y = y + I = (1+I) + (y - 1 + I) = w_o + v_o$. Therefore

(ii) $w_ny = (n + 1)w_n + v_o$.

Hence $w_n(y - n - 1) = v_o$ so that $V \subseteq w_nR$ for all n.

Let m be a non-zero element of M. We shall show that either $mR = V$ or $mR = w_nR$ for some n. Because every element of R is of the form $\Sigma a_{ij}x^iy^j$ we can write m in the form

(iii) $m = w_n a_n + \ldots + w_{n+j} a_{n+j} + v$ where $a_i \in C$ and $v \in V$ and $j$ is a non-negative integer. If $a_i = 0$ for all $i$ then $m$ is a non-zero element of $V$ so that $mR = V$. Suppose now that $m$ is as in (iii) with $a_n \neq 0$. By (ii) we have $m(y - n - j - 1) = w_n b_n + \ldots + w_{n+j-1} b_{n+j-1} + v'$ for some $b_i \in C$ and $v' \in V$ with $b_n = -ja_n \neq 0$. Repeating this process shows that $mR$ contains an element $m' = w_n c_n + t$ for some $c_n \in C$ and $t \in V$ with $c_n \neq 0$. But $m'y = (n + 1)w_n c_n + v_o c_n + ty$ by (ii) so that $t-(v_o c_n + ty)/(n+1) \in mR$. By writing $t$ as a linear combination of the $v_i$ it is easy to see that $t-(v_o c_n + ty)/(n + 1) \neq 0$. Therefore $mR$ contains a non-zero element of $V$ so that $V \subseteq mR$. Hence $t \in mR$ so that $w_n \in mR$. It now follows easily from (iii) that $mR = w_n R$. The argument given above can easily be modified to show that the $w_i$ are linearly independent mod($V$). For each non-negative integer $k$ we know that every element of $w_{k+1}R$ is of the form $w + v$ for some $v \in V$ where $w$ is a linear combination over $C$ of the $w_i$ with $i \geq k + 1$. Therefore $w_k \notin w_{k+1}R$.

Thus $0 \subseteq V = \bigcap_{n=o}^{\infty} w_n R \subseteq \ldots \subseteq w_n R \subseteq \ldots \subseteq w_1 R \subseteq w_o R = M$ is a complete list of the submodules of $M$. Clearly $M$ is a non-Artinian cyclic serial essential extension of the simple module $V$. We have $w_n x^i \in w_{n+1}R$ for all positive integers $i$. Also $w_n(y - n - 1) = v_o$ by (ii) so that $w_n y^j(y - n - 1) \in w_{n+1}R$ for all $j$. Therefore $w_n R/w_{n+1}R$ is annihilated by $y - n - 1$. It follows that if $i$ is a non-negative integer then $w_n R/w_{n+1}R$ is annihilated by $y - i - 1$ if and only if $i = n$, so that $w_n R/w_{n+1}R \cong w_i R/w_{i+1}R$ is and only if $i = n$. Also $w_n R/w_{n+1}R$ is not isomorphic to $V$ because $Vx \neq 0$. Thus the distinct simple subfactors of $M$ are non-isomorphic. □

Remarks

(1) To the best of our knowledge the first example of a critical non-compressible module over a left and right Noetherian ring was constructed by K.R. Goodearl [63]. In 7.15 the module M/V is 1-critical and not compressible.

(2) For further information about Noetherian fully bounded rings see the relevant papers by R. Gordon.

(3) It is not known whether a right and left Noetherian right fully bounded ring is always left fully bounded.

(4) It is not known whether a finitely-generated essential extension of an Artinian module over a left and right Noetherian ring of right Krull dimension 1 is always Artinian.

# 8 Semi–hereditary rings and p.p. rings

We shall characterise those semi-hereditary rings and p.p. rings which have semi-primary quotient rings. This will involve the use of weak chain conditions such as the non-existence of infinite sets of orthogonal idempotents. Some of the material comes from the work of R. Gordon and L.W. Small on piecewise domains, but we shall give a self-contained treatment. We shall also prove that hereditary Noetherian rings satisfy the restricted minimum condition.

As far as the theory of projective modules is concerned we assume only such basic information as that a module is projective if and only if it is a direct summand of a free module, and that if K is a submodule of M with M/K projective then K is a direct summand of M.

Let R be a ring and let n be a positive integer. By an "n-*generator right ideal*" we mean a right ideal which can be generated by n elements. We say that R is *right hereditary* (respectively *semi-hereditary*, n-*hereditary*, *p.p.*) if every (respectively finitely-generated, n-generator, principal) right ideal of R is projective.

Let R be any ring and let $x \in R$, then left multiplication by x gives a surjective homomorphism from R to xR whose kernel is $r(x)$. It follows easily that xR is projective if and only if $r(x) = eR$ for some idempotent e. Thus R is right p.p. if and only if for each $x \in R$ there is an idempotent e such that $r(x) = eR$.

Any integral domain is trivially a right p.p. ring. The ring Z[x] is a p.p. ring but is not semi-hereditary, and the ring of all algebraic integers is semi-hereditary but not hereditary. Examples of hereditary rings include Dedekind domains, full or triangular matrix rings over division rings, full matrix rings over Dedekind domains, and prime Noetherian rings such as 6.4. We have already shown that a left and right Noetherian p.p. ring is a direct sum of Artinian rings and prime rings (5.4).

Example 8.1. A triangular matrix ring over a hereditary ring need not be a p.p. ring. Let R be the ring of 2 by 2 upper triangular matrices over Z and set $x = 2e_{11} + e_{12}$. Each element of $r(x)$ is of the form $ze_{12} - 2ze_{22}$ for some $z \in Z$, and it follows easily that $r(x)$ contains no non-zero idempotents. But $r(x) \neq 0$, so that there is no idempotent e with $r(x) = eR$. □

Example 8.2. (Chase) The following ring R is left hereditary but not right p.p. Let T be the ring of all eventually-constant sequences of elements of Z/2Z with componentwise operations then T is a countable Boolean algebra. Let W be an ideal of T then W has a countable generating set $x_1, x_2, \ldots$ . By induction we may assume that there are orthogonal idempotents $e_1, \ldots, e_n$ (some of which may be zero) such that $x_1T + \ldots + x_nT = e_1T + \ldots + e_nT$. Set $e_{n+1} = (1 - e_1 - \ldots - e_n)x_{n+1}$ then $e_{n+1}$ is an idempotent which is orthogonal to $e_1, \ldots, e_n$ and we have $x_1T + \ldots + x_{n+1}T = e_1T + \ldots + e_{n+1}T$. Therefore there are orthogonal idempotents $e_1, e_2, \ldots$ such that $W = \Sigma e_iT$. This sum is direct so that W is projective. Therefore T is hereditary.

Let I be the ideal of T consisting of those sequences of elements of Z/2Z which are 0 from some point onwards then $T/I \cong Z/2Z$. Set $S = T/I$ and

$$R = \begin{bmatrix} S & S \\ 0 & T \end{bmatrix}.$$

Let $x = e_{12}$. Working in R we have $r(x) = e_{11}R \oplus Ie_{22}$. But the ideal I of T is not generated by an idempotent. Hence there is no idempotent e of R such that $r(x) = eR$ so that R is not right p.p. We note also that the right singular ideal of R is non-zero because $Se_{12} + Ie_{22}$ is an essential right ideal and $e_{12}(Se_{12} + Ie_{22}) = 0$ (cf. 8.3).

Now let L be a left ideal of R and set $A = \{t \in T: te_{22} \in L\}$. Then A is an ideal of T. Set $V = Se_{11} + Se_{12}$ then V is a vector space over S with $S \cong Z/2Z$. If $te_{22} \in L$ then $e_{12}te_{22} \in L$, i.e. $te_{12} \in L$. Thus $SAe_{12} \subseteq L$. Set $M = L \cap V$ and $B = SAe_{12}$ then M and B are S-subspaces of V with $B \subseteq M$. Therefore there is a subspace K of M such that $M = K \oplus B$. In fact K is a left ideal of R because $e_{12}V = 0 = e_{22}V$. There are only five possibilities for K, namely 0, $Se_{11}$, $Se_{12}$, $S(e_{11} + e_{12})$, and V, and it is easy to show that each of these is projective as a left R-module. For example, we have $Se_{11} = Re_{11}$ so that $Se_{11}$ is projective, and right multiplication by $e_{12}$ is an isomorphism from $Se_{11}$ to $Se_{12}$. Set $C = B + Ae_{22}$ then $C = RAe_{22}$ and $L = K \oplus C$. As in the first paragraph there are orthogonal idempotents $f_i \in T$ such that $A = \Sigma f_iT$. Thus $C = \Sigma Re_{22}f_i$ where the $e_{22}f_i$ are orthogonal idempotents. Each $Re_{22}f_i$ is projective and hence so also is C. Thus $L = K \oplus C$ with both K and C projective, so that L is projective. □

Lemma 8.3. Let R be a right p.p. ring then R is right non-singular.

Proof. Let $x \in R$ and suppose that $r(x)$ is an essential right ideal. We have $r(x) = eR$ for some idempotent e. Thus $r(x) \cap (1 - e)R = 0$ so that $1 - e = 0$. Therefore $x = 0$. □

When we say that a ring R has *no infinite sets of orthogonal idempotents* we mean that there do not exist infinitely-many non-zero idempotent elements $e_i$ of R such that $e_ie_j = 0$ whenever $i \neq j$.

Lemma 8.4. (Small) Let R be a right p.p. ring which has no infinite sets of orthogonal idempotents. Then R is a left p.p. ring, each right or left annihilator in R is generated by an idempotent, and R has the a.c.c. and d.c.c. for right annihilators.

Proof. Let A be a right annihilator in R and set $B = \ell(A)$. Suppose that B contains non-zero orthogonal idempotents $e_1,\ldots,e_n$ and set $e = e_1+\ldots+ e_n$ then e is idempotent. If B = Re we stop. If not, we have $B \neq Be$ so that $B(1 - e) \neq 0$. Let $y \in B(1 - e)$ with $y \neq 0$. Then $r(y) = (1 - f)R$ for some idempotent f. Because $B = \ell(A)$ and $A \subseteq (1 - f)R$ we have $f \in B$. Also $f \neq 0$ because $y \neq 0$. We have ye = 0 so that $e \in (1 - f)R$, i.e. $(1 - f)e = e$, i.e. fe = 0. Set $e_{n+1} = f - ef$ then $e_{n+1}$ is a non-zero idempotent element of B which is orthogonal to $e_1,\ldots,e_n$. Because R has no infinite sets of orthogonal idempotents this process must stop after a finite number of steps, and when it does we have B = Re for some idempotent e. We have $A = (1 - e)R$ with 1 - e idempotent.

Thus each right or left annihilator is generated by an idempotent. Let C and D be right annihilators in R with $C \subseteq D$. We have C = eR and D = fR for some idempotents e and f. Thus e = fe. Set g = f - ef then e and g are orthogonal idempotents and fR = eR + gR. Also g = 0 if and only if C = D. Because R has no infinite sets of orthogonal idempotents it follows that R has no infinite chains of distinct right annihilators. □

An idempotent of a ring R is called *primitive* if it is not the sum of two non-zero orthogonal idempotent elements of R. A ring R is said to have *enough idempotents* if the identity element of R can be written as the sum of a finite number of orthogonal primitive idempotent elements of R. A ring has enough idempotents if it has no infinite sets of orthogonal idempotents, but the converse is not true. For example, it was shown by J.C. Shepherdson that there is an integral domain R such that there are elements a and b of $M_2(R)$ with ab = 1 and ba $\neq$ 1. The ring $M_2(R)$ has enough idempotents because $e_{11}$ and $e_{22}$ are primitive, but the elements $b^n(1 - ba)a^n$ form an infinite set of orthogonal idempotents.

Lemma 8.5. Let R be a right p.p. ring with a primitive idempotent e. Let x, y $\in$ R with xey = 0 then xe = 0 or ey = 0.

Proof. Define f: eR $\to$ xeR by f(er) = xer for all r $\in$ R then f is a surjective homomorphism. We have ey $\in$ Ker(f). But xeR is projective so that Ker(f) is a direct summand of eR. Because e is primitive we know that eR is indecomposable. Therefore either Ker(f) = 0 in which case ey = 0, or Ker(f) = eR in which case xe = 0. □

Lemma 8.6. (Gordon and Small) Let R be a right or left p.p. ring with enough idempotents and let K be a nil right ideal of R then K is nilpotent.

Proof. There are orthogonal primitive idempotents $e_1, \ldots e_n$ such that $1 = e_1 + \ldots + e_n$. Let k $\in$ K then $ke_i$ is nilpotent for each i. But if $e_ike_i \neq 0$ then $e_i(ke_i)^2 = e_ike_ike_i \neq 0$ (8.5), and so on. Therefore $e_ike_i = 0$ so that $e_iKe_i = 0$. Hence $e_iK$ is nilpotent. Thus $e_1K + \ldots + e_nK$ is a sum of nilpotent right ideals and hence is nilpotent, and $K \subseteq e_1K + \ldots + e_nK$. □

Lemma 8.7. (Gordon and Small) Let R be a ring with an ideal I and a subring S such that $I \cap S = 0$ and R = I + S. Suppose also that R is right n-hereditary, then S is right n-hereditary.

Proof. Let K be an n-generator right ideal of S. Then KR is an n-generator right ideal of R and so is projective. Thus KR is a direct summand of a free right R-module F, so that KR/KRI is a direct summand of the free

R/I-module F/FI. Hence KR/KRI is projective as a right R/I-module, i.e. as a right S-module. But KRI = KI and KR = KS + KI = K + KI. Also $K \cap KI = 0$ because $S \cap I = 0$. Therefore $KR/KRI \cong K$ as right S-modules. □

We shall use 8.7 in the situation where R has an idempotent e such that eR is an ideal by taking I = eR and S = (1 - e)R(1 - e).

The next lemma will enable us to use induction in the proof of our main theorem about p.p. rings where we shall also use its notation. We shall write N(X) for the nilpotent radical of a ring X.

Lemma 8.8. Let R be a p.p. ring with no infinite sets of orthogonal idempotents. Then there is a non-zero idempotent element e of R with the following properties, where f = 1 - e, S = eRe, M = eRf, and T = fRf:

S and T are p.p. rings with no infinite sets of orthogonal idempotents and S is prime; fRe = 0; M is an ideal of R with $M^2 = 0$; R = S + M + T and the sum is direct; eR and Rf are ideals of R; $R/N(R) \cong S \oplus T/N(T)$.

Proof. It may be helpful to think of R as the matrix ring

$$\begin{bmatrix} S & M \\ 0 & T \end{bmatrix}.$$

By 8.4 there is a prime ideal P of R such that P is a left annihilator (cf. proof of 6.13). Hence P = R(1 - e) for some non-zero idempotent e of R (8.4). The rest of the proof follows from 8.7 and direct calculation (some of which was done in the proof of 6.13). □

Lemma 8.9. (Small) Let R be a right p.p. ring and let c be right regular mod(N) then c is right regular.

Proof. See proof of 5.4. □

Theorem 8.10. Let R be a right p.p. ring such that every right regular element of R is regular. Suppose also that R/N has finite right Goldie dimension where N is the nilpotent radical of R. Let c be a regular element of R then cN = N. Also $C(0) = C(N)$ and R/N is a right Goldie ring.

Proof. Because R/N has finite right Goldie dimension we know that R/N (and hence also R) has no infinite sets of orthogonal idempotents. Thus we can apply 8.8 and we shall use its notation. We know that S and T are p.p. rings (8.7), and that S and T/N(T) have finite right Goldie dimension because $R/N \cong S \oplus T/N(T)$. Also R has the a.c.c. for right annihilators (8.4) and hence so also does S. Thus S is a prime right Goldie ring. Let d be a right regular element of T then e + d is a right regular element of R and so is regular. Let $t \in T$ with td = 0 then t(e + d) = 0 so that t = 0. Thus T has the same properties as R.

The case where R is prime is easy because then R = S. Suppose now that R is not prime, i.e. that $T \neq 0$, then the right Goldie dimension of T/N(T) is strictly less than that of R/N. By induction we may suppose that T satisfies the conclusions of the theorem. Thus T/N(T) (and hence also R/N) is a right Goldie ring.

Now let a be a regular element of S. We shall show that aM = M. Let $x \in M$ and let g be a primitive idempotent in T. Set b = a + x + f - g then gb = 0. Thus b is not left regular in R and so is not right regular. Therefore $r(b) = hR$ for some non-zero idempotent h. Let h = s + m + t with $s \in S$ etc. Because bh = 0 and $h = h^2$ we have as = 0 so that s = 0; am + xt = 0; ft = gt; m = mt; and $t = t^2$. But $t \in fRf$ so that t = ft = gt. Therefore g - tg and tg are orthogonal idempotents in T. Because g is primitive we have either tg = g or tg = 0. If tg = 0 then $t = t^2 = (gt)^2 = 0$ and m = mt = 0 so that h = 0, which is a contradiction. Therefore tg = g.

Hence amg = -xtg = -xg so that xg $\in$ aM. But x = xf and f is a sum of primitive idempotents such as g. Therefore x $\in$ aM so that aM = M.

Forgetting the notation of the last paragraph, let c be a regular element of R then c = s + m + t for some s $\in$ S etc. We wish to show that t is right regular in T. Let z $\in$ T with tz = 0. Clearly s is regular in S so that sM = M. Hence sx = -mz for some x $\in$ M. We have c(x + z) = sx + mz = 0. Therefore z = 0 and t is right regular in T. We have N = M + N(T). But sM = M and tN(T) = N(T), so that cN = N. Also s is regular in S and t is regular mod(N(T)) from which it follows easily that c $\in$ $C(N)$. Therefore $C(0) \subseteq C(N)$, so that $C(0) = C(N)$ (8.9). □

A ring R is *semi-primary* if the Jacobson radical N of R is nilpotent and R/N is Artinian.

Theorem 8.11. Let R be a right p.p. ring in which every right regular element is regular and suppose that R/N has finite right Goldie dimension, then R has a semi-primary right quotient ring.

Proof. We know that $C(0) = C(N)$ and that R/N is a right Goldie ring (8.10). Let a $\in$ R and c $\in$ $C(0)$ then c $\in$ $C(N)$. By applying Goldie's theorem in R/N we have ad = cb + n for some d $\in$ $C(N)$, b $\in$ R, and n $\in$ N. But cN = N (8.10) so that n = cx for some x $\in$ N. Also d $\in$ $C(0)$. Thus ad = c(b + x) so that R satisfies the right Ore condition with respect to $C(0)$. Therefore R has a right quotient ring Q.

For each c $\in$ $C(0)$ we have cN = N so that $N = c^{-1}N$. Therefore QN = N so that NQ is a nilpotent ideal of Q. Let S be the right quotient ring of R/N and define f:Q → S by $f(ac^{-1}) = (a + N)(c + N)^{-1}$ for all a $\in$ R and c $\in$ $C(0)$. Then f is a well-defined surjective ring homomorphism. Also Ker(f) = NQ so that Q/NQ ≅ S. Therefore Q/NQ is Artinian. □

A ring R is said to have the *right essentiality condition* if each right regular element of R generates an essential right ideal. This condition and the condition that right regular elements are regular both hold in commutative rings and in right non-singular rings of finite right Goldie dimension (1.11 and 1.12). If R is an integral domain which does not have right Goldie dimension 1 then R is a p.p. ring and right regular elements of R are regular, but R does not have the right essentiality condition. If R is a ring which has a semi-primary right quotient ring Q and c is right regular in R then c is right regular in Q and hence is a unit in Q, so that c is regular and cR is essential in R.

The next main result will show that certain right semi-hereditary rings have semi-primary right quotient rings, but this time we will not assume that R/N has finite right Goldie dimension.

<u>Lemma 8.12</u>. Let R be a right 2-hereditary ring with enough idempotents and with the right essentiality condition, then R/N has finite right Goldie dimension.

<u>Proof</u>. There are orthogonal primitive idempotents $e_1,\dots,e_n$ of R such that $R = e_1R \oplus \dots \oplus e_nR$. We shall show that each $e_iR/e_iN$ is a uniform module, from which it follows that R/N is a direct sum of uniform right ideals.

Let e be a primitive idempotent in R. We know that N is nilpotent (8.6), and that if $x, y \in R$ with $xey = 0$ then $xe = 0$ or $ey = 0$ (8.5). Let A be a non-nilpotent submodule of eR then $A^2$ is not contained in N so that $ae \notin N$ for some $a \in A$. If $r \in R$ with $aer = 0$ then $er = 0$. Set $c = ae + 1 - e$ then it follows that c is right regular and hence that cR is an essential right ideal of R. Therefore $cR \cap eR$ is essential in eR. But $cR \cap eR \subseteq aeR$ so that aeR (and hence also A) is essential in eR.

Now let A and B be non-nilpotent submodules of eR. As before we have $ae \notin N$ and $be \notin N$ for some $a \in A$ and $b \in B$. Define $f:(aeR \oplus beR) \to (aeR + beR)$ by $f((aex,bey)) = aex + bey$ for all $x, y \in R$ then $Ker(f) = \{(z,-z) : z \in aeR \cap beR\}$. Thus $Ker(f) \cong aeR \cap beR$. But Ker(f) is a direct summand of $aeR \oplus beR$ because $aeR + beR$ is projective. Therefore $aeR \cap beR$ is a homomorphic image of $aeR \oplus beR$. But $aeR \oplus beR$ is generated by (ae,0) and (0,be), so that $aeR \cap beR$ is generated by elements u and v such that ue = u and ve = v. Because $ae \notin N$ and $be \notin N$ it follows from the last paragraph that aeR and beR are essential in eR so that $aeR \cap beR \neq 0$. Thus one of u or v is non-zero, say $u \neq 0$. We have $eu = u = ue \neq 0$. Therefore $u^2 = ueu \neq 0$ (8.5), and so on. Continuing this process shows that u is not nilpotent so that $u \notin N$. Therefore $A \cap B$ is not contained in N, so that eR/eN is uniform. □

Theorem 8.13. (Gordon) Let R be a right 2-hereditary ring with enough idempotents and with the right essentiality condition, then R has a semi-primary right quotient ring.

Proof. Combine 8.12 and 8.11. □

In 8.13 we did not need to assume that R/N has finite right Goldie dimension because we could prove it, but the following example shows that it was necessary to assume it for the p.p. rings in 8.11.

Example 8.14. (Gordon and Small) Let S be the opposite ring of 1.22 then S is an integral domain which has a left quotient division ring D but S has infinite right Goldie dimension. Let

$$R = \begin{bmatrix} S & D \\ 0 & D \end{bmatrix}.$$

Then R has right Goldie dimension 2 because $e_{11}R$ and $e_{22}R$ are uniform. Because S is a direct summand of R/N we know that R/N does not have finite right Goldie dimension. It can easily be shown by direct calculation that R is right p.p. For example, if $s \in S$ and $d \in D$ with $s \neq 0$ then $r(se_{11} + de_{12}) = eR$ where e is the idempotent $-s^{-1}de_{12} + e_{22}$. □

We conclude the chapter with some assorted useful results.

Theorem 8.15. (Warfield) Let R be a right non-singular right serial ring then R is right semi-hereditary.

Proof. There are orthogonal idempotents $e_1,\ldots,e_n$ such that $R=e_1R \oplus\ldots\oplus e_nR$ and each $e_iR$ is serial. Let e be one of the $e_i$ and let K be a non-zero finitely-generated submodule of eR. Because K is a finite sum of cyclic modules and eR is serial we have $K = exR$ for some $x \in R$. Also $exR = exe_1R + \ldots + exe_nR$ so that $K = exe_iR$ for some i, say $K = exe_1R$. Define $f:e_1R \to K$ by $f(e_1r) = exe_1r$ for all $r \in R$ then $\text{Ker}(f) \neq e_1R$. Because $e_1R$ is uniform and R is right non-singular it follows that $\text{Ker}(f) = 0$. Therefore $e_1R \cong exe_1R$ so that K is projective.

Now let A be a non-zero finitely-generated right ideal of R. We wish to show that A is projective, and because R has finite right Goldie dimension we may suppose that A is indecomposable. Let e be one of the $e_i$ such that $eA \neq 0$ then eA is finitely-generated and hence projective. Because A is indecomposable it follows that $A \cong eA$. □

Lemma 8.16. Let R be a right n-hereditary ring where n is a positive integer, let F be a free right R-module of rank n, and let K be a n-generator submodule of F, then K is projective.

Proof. Fix a basis of n elements for F and let $f_1,\ldots,f_n$ be the corresponding co-ordinate maps from F to R (i.e. $f_i(a_1,\ldots,a_n) = a_i$). Set $A_1 = f_1(K)$ then $A_1$ is an n-generator right ideal of R and so is projective. Therefore there are submodules $K_1$ and $B_1$ of K such that $K = K_1 \oplus B_1$ with $f_1(K_1) = 0$ and $B_1 \cong A_1$. Similarly there are submodules $K_2$ and $B_2$ of $K_1$ such that $K_1 = K_2 \oplus B_2$ with $f_1(K_2) = f_2(K_2) = 0$ and $B_2$ projective. Eventually we obtain $K = B_1 \oplus \ldots \oplus B_n$ with each $B_i$ projective. □

Theorem 8.17. (Small) Let R be any ring and let n be a positive integer, then R is right n-hereditary if and only if $M_n(R)$ is right p.p.

Proof. Let F be a free right R-module of rank n and let S be its endomorphism ring. We must show that R is right n-hereditary if and only if S is right p.p. Throughout this proof right annihilators will be taken in S.

Suppose that S is right p.p. and let I be an n-generator right ideal of R. There is a submodule K of F such that $K \cong I$. Because K has n generators there exists $s \in S$ such that $sF = K$ (we regard F as being a left S-module in the usual way). Clearly $r(s)F \subseteq \text{Ker}(s)$. Let $x \in \text{Ker}(s)$. Then $xR = tF$ for some $t \in S$. Thus $stF = 0$ so that $st = 0$. It follows that $\text{Ker}(s) = r(s)F$. But $r(s) = eS$ for some idempotent e. Hence $\text{Ker}(s) = eF$ so that Ker(s) is a direct summand of F. Therefore K is projective.

Conversely suppose that R is right n-hereditary and let $s \in S$. Because sF is an n-generator submodule of F we know that sF is projective (8.16). Therefore Ker(s) is a direct summand of F, i.e. $\text{Ker}(s) = eF$ for some idempotent $e \in S$. Hence $r(s) = eS$. □

Corollary 8.18. Let R be a left and right Noetherian right hereditary ring then R is left hereditary.

Proof. For each positive integer n we know that $M_n(R)$ is right p.p. (8.17).

Also $M_n(R)$ is Noetherian and so has no infinite sets of orthogonal idempotents. Therefore $M_n(R)$ is left p.p. (8.4). Thus R is left n-hereditary (8.17). But every left ideal of R has n generators for some positive integer n. Therefore R is left hereditary. □

Corollary 8.19. (Small) Let R be a right semi-hereditary ring such that $M_n(R)$ has no infinite sets of orthogonal idempotents for each positive integer n, then R is left semi-hereditary.

Proof. Modify the proof of 8.18. □

The conditions of 8.19 are satisfied for example if R is a right semi-hereditary ring which has finite right Goldie dimension or which is a subring of a semi-primary ring.

Lemma 8.20. (Dual basis lemma) Let M be a right R-module. Then M is projective if and only if M has a dual basis, i.e. if and only if there are elements $x_i$ of M and corresponding homomorphisms $f_i:M \to R$ such that if $x \in M$ then (i) $f_i(x) = 0$ for all but finitely-many i, and (ii) $x = \sum_i x_i f_i(x)$.

Proof. Suppose that M is projective. Then M is a direct summand of a free module F. Fix a basis $y_i$ of F and let $g_i:F \to R$ be the corresponding coordinate maps. For each i let $x_i$ be the component of $y_i$ in M and let $f_i$ be the restriction of $g_i$ to M, then $\{(x_i,f_i)\}$ is a dual basis for M.

Conversely, suppose that $\{(x_i,f_i)\}$ is a dual basis for M. Let F be the free module with a basis consisting of elements $y_i$ which are in one-to-one correspondence with the $x_i$. Define $p:F \to M$ by $p(y_i) = x_i$ for all i, and define $f:M \to F$ by $f(x) = \sum_i y_i f_i(x)$ for all $x \in M$. It follows from (i) and (ii) above that f is well-defined. Clearly pf is the identity function on M so that M is isomorphic to a direct summand of F and so is projective. □

We recall that a ring $R$ is said to have the right restricted minimum condition if $R/K$ is Artinian for each essential right ideal $K$ of $R$.

Theorem 8.21. Let $R$ be a left and right Noetherian hereditary ring then $R$ has the left and right restricted minimum conditions.

Proof. Let $I$ be a right ideal of $R$ and let $I^*$ be the set of all right $R$-module homomorphisms from $I$ into $R$, then $I^*$ is in a natural way a left $R$-module. Because $I$ is finitely-generated and projective, there exist $x_1,\ldots,x_n \in I$ and $f_1,\ldots,f_n \in I^*$ such that $x = x_1f_1(x) + \ldots + x_nf_n(x)$ for all $x \in I$ (8.20). Let $g \in I^*$ then $g(x) = \sum_i g(x_i)f_i(x)$ for all $x \in I$ so that $g = \sum_i g(x_i)f_i$ with $g(x_i) \in R$. Thus $I^*$ is finitely-generated as a left $R$-module by the elements $f_1,\ldots,f_n$.

Let $A$ and $B$ be right ideals of $R$ such that $B$ is an essential submodule of $A$. Let $f$ and $g$ belong to $A^*$ and assume that $f$ and $g$ have the same restriction to $B$. Let $a \in A$. Then $aK \subseteq B$ for some essential right ideal $K$ of $R$ (1.1). For each $k \in K$ we have $f(a)k = f(ak) = g(ak) = g(a)k$. Hence $(f(a) - g(a))K=0$. Therefore $f(a) = g(a)$ (8.3) so that $f = g$. Thus by identifying each element of $A^*$ with its restriction to $B$ we can consider $A^*$ to be a subset of $B^*$. Suppose that $A^* = B^*$. Let $\{(b_1,g_1),\ldots,(b_n,g_n)\}$ form a dual basis for $B$. There exist $f_i \in A^*$ such that $g_i$ is the restriction of $f_i$ to $B$. With $a$ and $K$ as above we have $ak = \sum_i b_ig_i(ak) = (\sum_i b_if_i(a))k$ so that $a = \sum_i b_if_i(a)$. Therefore $a \in B$ so that $A = B$.

Now let $K \subseteq \ldots \subseteq A_n \subseteq \ldots \subseteq A_1$ be essential right ideals of $R$. Considering each $A_n{}^*$ to be a subset of $K^*$ as above, we have $A_1{}^* \subseteq \ldots \subseteq A_n{}^* \subseteq \ldots \subseteq K^*$. But $K^*$ is a finitely-generated left $R$-module and $R$ is left Noetherian. Therefore the $A_n{}^*$ and hence also the $A_n$ are equal from some point onwards. □

Theorem 8.22. Let R be a left and right Noetherian hereditary ring then R is a direct sum of Artinian rings and prime rings.

Proof. This is a special case of 5.4 but the following is the sketch of a more direct proof. Suppose that R is indecomposable and not prime. Let P be a prime ideal of R then P is not generated by a central idempotent. Therefore there are no idempotents e and f such that P = eR = Rf. Hence either P is not a right annihilator (8.4) in which case P is an essential right ideal of R so that R/P is Artinian (8.21), or P is not a left annihilator and again R/P is Artinian. Thus every prime factor ring of R is Artinian, so that R/N (and hence also R) is Artinian. □

Proposition 8.23. (Levy) Let R be a semi-prime p.p. ring with no infinite sets of orthogonal idempotents then R is a direct sum of prime rings.

Proof. In the notation of 8.8 we have M = 0 so that $R = S \oplus T$. The result follows by induction on the number of minimal primes (8.4 and 1.16). □

Proposition 8.24. (Sandomierski) Let P be a non-singular projective right R-module and suppose that P has a finitely-generated essential submodule K, then P is finitely-generated.

Proof. Let $\{(x_i, f_i)\}$ be a dual basis for P (8.20) and let S be a finite generating set for K. For each $s \in S$ we have $f_i(s) = 0$ for all but finitely-many values of i. Thus there are only finitely-many of the $f_i$, say $f_1, \ldots, f_n$, which do not act as the zero function on K. Let $x \in P$ then there is an essential right ideal L of R such that $xL \subseteq K$ (1.1). For each $y \in L$ we have $xy = \sum_i x_i f_i(xy) = x_1 f_1(xy) + \ldots + x_n f_n(xy)$. Thus $(x - x_1 f_1(x) - \ldots - x_n f_n(x))L = 0$ so that $x = x_1 f_1(x) + \ldots + x_n f_n(x)$. Therefore $x_1, \ldots, x_n$ form a finite generating set for P. □

Corollary 8.25. (Sandomierski) Let R be a right hereditary ring of finite right Goldie dimension then R is right Noetherian.

Proof. Let P be a right ideal of R then there are uniform submodules $U_1,\ldots,U_n$ of P such that the sum $U_1 + \ldots + U_n$ is direct and essential in P (1.9(a)). Replacing each $U_i$ by a non-zero cyclic submodule we may suppose that each $U_i$ is cyclic. Therefore P has a finitely-generated essential submodule so that P is finitely-generated (8.24). □

Remarks.

(1) Results proved in [32] include that if R is right hereditary or semi-hereditary or p.p. then so also is eRe for any idempotent $e \in R$, and that if R is any ring then R is right hereditary if and only if the endomorphism ring of every free right R-module is right p.p.

(2) It was shown by B.L. Osofsky that a right hereditary right self-injective ring is semi-simple Artinian [112].

(3) It was shown by J. Fuelberth, E. Kirkman and J. Kuzmanovich [50] that if R is a left and right hereditary ring which is finitely-generated as a module over its centre then R is the direct sum of an Artinian ring and a semi-prime ring.

(4) In [89] S. Jondrup has shown using 8.11 that if R is a semi-prime right p.p. ring which satisfies a polynomial identity and which is finitely-generated as an algebra over its centre then R has a right quotient ring (Cf. Chapter 1, Remark (6)), and if in addition R is finitely-generated as a module over its centre then the right quotient ring of R is regular and is obtained by inverting the regular central elements of R.

(5) The restricted minimum condition for semi-prime hereditary Noetherian rings was proved by D.B. Webber in [157].

# 9 Orders in semi–primary rings

In Chapters 4 and 5 we proved some results about Noetherian orders in Artinian rings by regarding such rings as being special Noetherian rings. By contrast we shall now study general orders in semi-primary rings of which Noetherian orders in Artinian rings are a special case. There will also be a connection with Chapter 8 because we shall prove the result of J. Fuelberth and J. Kuzmanovich that a left and right hereditary order in a semi-primary ring is a direct sum of semi-primary rings and prime rings. The material of this chapter is taken from a paper by A.W. Chatters and J.C. Robson.

We recall that a ring R is semi-primary if the Jacobson radical N of R is nilpotent and R/N is Artinian. To say that R is an order in a ring Q means that Q is the left and right quotient ring of R.

<u>Example 9.1</u>. Let K be any subfield of the complex numbers, set

$$R = \begin{bmatrix} Z & K \\ 0 & Z \end{bmatrix} \quad \text{and} \quad S = \begin{bmatrix} Q & K \\ 0 & Q \end{bmatrix} .$$

Then R is an order in the semi-primary ring S. □

<u>Lemma 9.2</u>. Let R be an order in the semi-primary ring Q, let N be the largest nil ideal of R, and let N' be the Jacobson radical of Q. Then N' = NQ = QN and N is nilpotent. Also R/N is a semi-prime Goldie ring and $\mathcal{C}(0) = \mathcal{C}(N)$ in R.

<u>Proof</u>. Set $K = R \cap N'$ then K is a nilpotent ideal of R so that $K \subseteq N$. Working in R let $c \in \mathcal{C}(0)$ and suppose that $cr \in K$ for some $r \in R$. We have

$cr \in N'$ and $c$ is a unit of $Q$. Therefore $r \in N'$ so that $r \in K$. Thus $C(0) \subseteq C(K)$. Now suppose that $c \in C(K)$ and that $cq \in N'$ for some $q \in Q$. We have $q = rd^{-1}$ for some $r \in R$ and $d \in C(0)$. Thus $cqd \in N' \cap R$, i.e. $cr \in K$. Hence $r \in K$ so that $q \in N'$ (because $N' = KQ$). Thus $c + N'$ is regular in $Q/N'$. Hence $c$ is a unit of $Q$ so that $c$ is regular in $R$.

Thus $C(0) = C(K)$. It now follows easily that $Q/N'$ is the quotient ring of $R/K$ so that $R/K$ is a semi-prime Goldie ring (1.27). Therefore $R/K$ has no non-zero nil ideals (1.7). Thus $K = N$. □

Notation: Throughout this chapter $A$ is the sum of all the Artinian right ideals of $R$, $e$ is an idempotent in $R$ such that $e + N$ is a central idempotent which generates the socle of $R/N$, and $H = \cap Rc$ where $c$ ranges over the regular elements of $R$. We shall use $M_R$ to indicate that $M$ is being considered as a right $R$-module.

Lemma 9.3. Let $R$ be an order in the semi-primary ring $Q$ and let $M$ be a right $Q$-module such that $MN = 0$ and $M_R$ is finitely-generated, then $M_R$ is Artinian and $M = Me$.

Proof. This is really part of the proof of 5.2. □

Corollary 9.4. Let $R$ be an order in the semi-primary ring $Q$ and let $M$ be a right $Q$-module such that $M_R$ is Noetherian, then $M_R$ is Artinian and $M = Me$.

Proof. Because $NQ = QN$ (9.2) we have $MNQ = MQN = MN$. Similarly $MN^i$ is a right $Q$-module for each $i$. We can now apply 9.3 to $MN^i/MN^{i+1}$. □

Lemma 9.5. Let $R$ be an order in the semi-primary ring $Q$ then $e \in H$ and $H = HQ$.

Proof. See the proof of 5.1. □

Lemma 9.6. Let R be an order in the semi-primary ring Q and suppose that R is semi-hereditary, then $N = NQ = QN$.

Proof. We could derive this from 8.10, but the following proof due to R.B. Warfield is more direct. We must show that $cN = N$ for each regular element c of R, and because N is nilpotent it is enough to show that $N = cN + N^2$. Let $x \in N$ and set $I = xR + cR$. Because $I_R$ is projective it follows that $(I/IN)_{R/N}$ is projective and hence torsion-free. Thus if $y \in I$ and $yd \in IN$ for some $d \in \mathcal{C}(N)$ then $y \in IN$. We have $\mathcal{C}(0) = \mathcal{C}(N)$ (9.2). Because $c^{-1}x \in QN$ and $QN = NQ$ we have $c^{-1}x = nd^{-1}$ for some $n \in N$ and $d \in \mathcal{C}(0)$. Thus $xd = cn$ so that $xd \in IN$. Therefore $x \in IN$. But $IN = cN + xN$ and $x \in N$. Hence $x \in cN + N^2$. □

Theorem 9.7. (Fuelberth and Kuzmanovich) Let R be a left and right hereditary order in the semi-primary ring Q then R is a direct sum of semi-primary rings and prime rings.

Proof. Suppose that R is indecomposable. Set $T = Q/NQ$ then T is the quotient ring of R/N (cf. proof of 9.2). Let I be a minimal right ideal of T then $I \cap (R/N)$ is a uniform right ideal of R/N so that I has a cyclic essential R/N-submodule.

Let B be a right ideal of R such that $BQ = B$. We have $BNQ = BN$ (9.6). Set $M = B/BN$ then M is a right T-module. Also $M_{R/N}$ is projective because $B_R$ is projective. Let K be a simple direct summand of $M_T$, then $K_{R/N}$ is projective. Also K is isomorphic to a minimal right ideal of T so that K has a cyclic essential R/N-submodule. Therefore $K_{R/N}$ is finitely-generated (8.24). Hence $K = Ke$ (9.3). But $M_T$ is a direct sum of modules like K. Therefore $M = Me$, i.e. $B = Be + BN$. But $HQ = H$ (9.5), so that we can take $B = HN^i$ for any non-negative integer i. Hence $HN^i = HN^ie + HN^{i+1}$ for all i.

Because N is nilpotent it follows that H = He. But $e \in H$ (9.5) so that H = Re. Also H is a right ideal of R (9.5). Therefore Re = ReR. By symmetry we have eR = ReR = Re from which it follows that e is central.

We are assuming that R is indecomposable, so that either e = 1 or e = 0. If e = 1 then R = H so that R = Rc for all $c \in C(0)$ and hence R = Q. Suppose now that e = 0 then H = 0. Let $c \in C(0)$ then $Nc^{-1} \subseteq N$ (9.6) so that N = Nc. Thus $N \subseteq Rc$ so that $N \subseteq H$. Therefore N = 0, and it follows from 8.23 that R is prime. □

The ring R in 4.8 is a right (but not left) hereditary ring which is indecomposable and neither Artinian nor prime, and R is an order in an Artinian ring.

<u>Lemma 9.8</u>. Let R be an order in the semi-primary ring Q and let I be an Artinian right ideal of R, then I = IQ = Ie and $I_R$ has finite length.

<u>Proof</u>. We first note that if X is a right ideal of R with X = Xe then by 9.5 we have X = Xe = XH = XHQ = XQ. Suppose that $I \neq Ie$. Because I is Artinian we may suppose that I is a "minimal counter-example", i.e. that K = Ke for each proper submodule K of I. Clearly I is not the sum of its proper submodules, so that I has a unique maximal submodule M. We have M = Me = MQ. If $x \in I$ and $c \in C(0)$ with $xc \in M$ then, because $x = xcc^{-1}$, we have $x \in MQ$, i.e. $x \in M$. Therefore I/M is a simple torsion-free right R/N-module and so is isomorphic to a minimal right ideal of R/N. But the socle of R/N is generated by the central idempotent e + N. Hence (I/M)e=I/M, so that I = Ie + M = Ie + Me = Ie.

Thus I = Ie = IQ for each Artinian right ideal I of R. Hence each R-submodule of I is also a Q-submodule so that $I_Q$ is Artinian. Thus for each i we know that $IN^i/IN^{i+1}$ is an Artinian module over the semi-simple Artinian ring $Q/NQ$. Therefore each $IN^i/IN^{i+1}$ has finite length as a Q-module and hence also as an R-module. □

Corollary 9.9. Let R be an order in the semi-primary ring Q then $A = AQ = Ae \subseteq Re \subseteq H = HQ$.

Proof. Combine 9.8 and 9.5. □

Theorem 9.10. Let R be a left and right Noetherian ring which has an Artinian quotient ring. Then there is a central idempotent element e of R such that

(i) (Ginn and Moss) eR is the Artinian radical of R,

(ii) $(eR + N)/N$ is the socle of $R/N$, and

(iii) $eR = \cap cR = \cap Rc$ where c ranges over the regular elements of R.

(This is the same as 5.1.)

Proof. Because $H = HQ$ (9.5) and $H_R$ is Noetherian we know that $H_R$ is Artinian (9.4). Thus $H \subseteq A$. Therefore $A = Re = ReR = H$ (9.9) and the result follows by symmetry. □

Let R be an order in a semi-primary ring Q. The left and right Artinian radicals of R need not be equal (4.8). The following example shows that in general $\cap cR \neq \cap Rc$, and also that it is possible to have an ideal I of R with $I = IQ$ but I is not an ideal of Q (cf. 1.31).

Example 9.11. In this example Q will denote the field of rational numbers. Let W be the set of all formal series of the form $\sum_{i=k}^{\infty} a_i x^i$ with $k \in Z$ and $a_i \in Q$. Let V be the subset consisting of those elements of W for which $a_i \in Z$ whenever i is negative. We can regard V as being a $Z[x]$-module, and by writing rational functions as Laurent series we can make W into a $Q(x)$-module. Set

$$R = \begin{bmatrix} Z[x] & V \\ 0 & Z \end{bmatrix} \quad \text{and} \quad S = \begin{bmatrix} Q(x) & W \\ 0 & Q \end{bmatrix}$$

then R is a subring of S and S is semi-primary. Regular elements of R have non-zero diagonal entries and so are units of S. We know that Q(x) is the quotient ring of Z[x]. Also if $w \in W$ then $zw \in V$ for some $z \in Z$ with $z \neq 0$. It is now straightforward to show that if $s \in S$ then there is a regular element c of R such that $cs \in R$ and $sc \in R$ (in fact c can be taken to be a diagonal matrix). Thus S is the quotient ring of R. Let U be the ring of power series in x over Q and set $I = Ue_{12}$. Then I is an ideal of R and $IS = I$. We have $U = \cap zV$ where z ranges over the non-zero elements of Z, so that I = H in our standard notation. Because $x^{-1}U$ is not contained in U it follows that $SI \neq I$. Since $x^n e_{11} + e_{22}$ is a regular element of R for each positive integer n it is easy to show that $\cap cR = 0$ where c ranges over the regular elements of R. Therefore the only ideal B of R such that SB = B is B = 0. □

# 10 Rings with finite global dimension

It is well-known that a commutative Noetherian local ring of finite global dimension is an integral domain. In the non-commutative case M. Ramras asked whether a right Noetherian local ring R of finite right global dimension is always a matrix ring over an integral domain. In this chapter we shall use only the rank function of Chapter 2 and elementary properties of projective modules to show that in these circumstances R/N is a matrix ring over a domain, and then we shall give an example due to J.T. Stafford which shows that R need not be a semi-prime ring. It is known that R is a matrix ring over a domain in the special cases where R has a right Artinian right quotient ring or R is right non-singular; these results were proved by M. Ramras and R. Walker respectively and we give them in 10.7 and 10.8. If R is also assumed to be left Noetherian then the answer to Ramras's question is not known, but positive results have been obtained in some special cases and we shall consider one of these in Chapter 12.

We will use J and N to denote the Jacobson and nilpotent radicals respectively of a ring R and we shall say that R is *local* if R/J is a simple Artinian ring. A ring R has finite right global dimension if there is an integer n such that for each right R-module M there is an exact sequence $0 \to P_n \to P_{n-1} \to \ldots \to P_1 \to P_0 \to M \to 0$ where $P_i$ is a projective right R-module (some of the $P_i$ may be zero); the smallest such n is called the *right global dimension* of R and we write rgd(R) = n. Whenever we write rgd(R) = n it will be understood that n is an integer.

We have rgd(R) = 0 if and only if R is semi-simple Artinian, and rgd(R) =1 if and only if R is right hereditary and not semi-simple Artinian. If $rgd(R) = n$ then $rgd(M_k(R)) = n$, and $rgd(R[x]) = n + 1$ (by the Hilbert syzygy theorem). In general, the global dimensions of a ring and its factor rings are unrelated, e.g. $rgd(Z) = 1$ but $Z/4Z$ does not have finite global dimension.

Lemma 10.1. (Nakayama) Let R be any ring and let M be a finitely-generated right R-module with $M = MJ$ then $M = 0$.

Proof. Let $x_1,\ldots,x_n$ be a finite generating set for M. We have $x_1 = x_1 j_1 + \ldots + x_n j_n$ for some $j_i \in J$. Thus $x_1(1-j_1) \in x_2R + \ldots + x_nR$ and $1 - j_1$ is a unit of R. It follows that $x_1 \in x_2R + \ldots + x_nR$ so that $x_2,\ldots,x_n$ is also a generating set for M, and so on. □

Lemma 10.2. Let R be any ring and let P and Q be finitely-generated projective right R-modules such that $P/PJ$ is a direct summand of $Q/QJ$ as right $R/J$-modules, then P is a direct summand of Q as right R-modules.

Proof. Let $f:Q \to P/PJ$ be the composite of the natural map $Q \to Q/QJ$ with the projection $Q/QJ \to P/PJ$ and let $g:P \to P/PJ$ be the natural map, then, because Q is projective, there exists $h:Q \to P$ such that $gh = f$. Because gh is surjective we have $P = \mathrm{Im}(h) + \mathrm{Ker}(g) = \mathrm{Im}(h) + PJ$. It follows from Nakayama's lemma applied to the module $P/\mathrm{Im}(h)$ that $P = \mathrm{Im}(h)$. Thus $h:Q \to P$ is surjective with P projective, so that P is (isomorphic to) a direct summand of Q. □

Corollary 10.3. Let R be a local ring then there is (up to isomorphism) a unique finitely-generated indecomposable projective right R-module.

Proof. Because $R/J$ is simple Artinian, every finitely-generated right $R/J$-module is the direct sum of a finite number of copies of the unique simple right R/J-module. Hence if A and B are finitely-generated right R/J-modules then A is a direct summand of B or vice versa. It follows that if P and Q are finitely-generated projective right R-modules then P is a direct summand of Q or vice versa (10.2). □

There are many rings (e.g. Z) which are not local but have a unique finitely-generated projective module.

The following result is based on an idea of R. Walker. We recall that $\rho(M)$ denotes the rank of a right R-module M (cf. Chapter 2).

Lemma 10.4. Let R be a right Noetherian ring of finite right global dimension and suppose that there is a unique finitely-generated indecomposable projective right R-module P. Let M be any finitely-generated right R-module then $\rho(P)$ divides $\rho(M)$.

Proof. Let Q be any finitely-generated projective right R-module then Q is a direct sum of indecomposable modules so that $Q \cong P^k$ for some integer k. Hence $\rho(P)$ divides $\rho(Q)$. There is an exact sequence $0 \to P_n \to P_{n-1} \to \dots \to P_1 \to P_0 \to M \to 0$ where each $P_i$ is finitely-generated projective. Therefore, because the rank function is additive, we have $\rho(M) = \rho(P_0) - \rho(P_1) + \dots + (-1)^n \rho(P_n)$. But $\rho(P)$ divides $\rho(P_i)$ for all i. Therefore $\rho(P)$ divides $\rho(M)$. □

Lemma 10.5. (a) Let R be any ring and let U be a non-singular uniform right R-module then the endomorphism ring of U is an integral domain. (b) Let R be a right non-singular ring which is the direct sum of a finite number of isomorphic uniform right ideals, then R is isomorphic to a matrix ring over an integral domain.

Proof. (a) It is enough to show that every non-zero endomorphism of U is injective. Let f be an endomorphism of U with $\mathrm{Ker}(f) \neq 0$, then $\mathrm{Ker}(f)$ is an essential submodule of U. Let $u \in U$ then $uL \subseteq \mathrm{Ker}(f)$ for some essential right ideal L of R (1.1). Thus $f(u)L = f(uL) = 0$. Because U is non-singular we have $f(u) = 0$ so that $f = 0$.

(b) This follows from (a) together with the fact that if A and B are modules with $A \cong B^k$ for some positive integer k then the endomorphism ring of A is isomorphic to $M_k(S)$ where S is the endomorphism ring of B. □

<u>Theorem 10.6</u>. Let R be a right Noetherian ring of finite right global dimension and suppose that there is a unique finitely-generated indecomposable projective right R-module P, then there is a positive integer k such that $R \cong M_k(S)$ where S is a right Noetherian ring of finite right global dimension such that every finitely-generated projective right S-module is free and $S/N(S)$ is an integral domain. In particular R/N is a matrix ring over a domain. Also N, considered as a right R-module, is torsion with respect to C(N).

<u>Proof</u>. Set $d = \rho(P)$ then d divides $\rho(M)$ for each finitely-generated right R-module M (10.4). Let e be a primitive idempotent element of R then $eR \cong P$ so that $\rho(eR) = d$. Let K be any submodule of eR then $\rho(eR) = \rho(K) + \rho(eR/K)$ so that $\rho(K) \leq \rho(eR)$. But $\rho(eR)$ divides $\rho(K)$. Therefore either $\rho(K) = 0$ or $\rho(K) = d$. If $\rho(K)=0$ then for each $k \in K$ we have $kc=0$ for some $c \in C(N)$ from which it follows that $K \subseteq eN$. On the other hand, if $\rho(K) = d$ then $\rho(eR/K) = 0$ so that $ec \in K$ for some $c \in C(N)$ from which it follows easily that K is not contained in eN.

Let A and B be submodules of eR which are not contained in eN, define $f:(A \oplus B) \to (A + B)$ by $f((a,b)) = a + b$. Then $\mathrm{Ker}(f) \cong A \cap B$. Hence $\rho(A \oplus B) = \rho(A \cap B) + \rho(A + B)$. But $\rho(A) = \rho(B) = \rho(A + B) = d$. Therefore

$\rho(A \cap B) = d$ so that $A \cap B$ is not contained in eN. Hence eR/eN is a uniform module.

There are orthogonal primitive idempotents $e_1, \ldots, e_k$ such that $R = e_1R \oplus \ldots \oplus e_kR$. For each i we have $\rho(e_iN) = 0$ from above. Also $N = e_1N \oplus \ldots \oplus e_kN$. Therefore $\rho(N) = 0$, i.e. N is torsion as a right R-module with respect to C(N). For each i we have $e_iR \cong P$ so that $e_iR \cong e_jR$ for all i and j. Hence R/N is the direct sum of the isomorphic right ideals $e_iR/e_iN$ so that R/N is a matrix ring over a domain (10.5(b) and 1.6). Let S be the endomorphism ring of P then S is isomorphic to the endomorphism ring of $e_iR$ for each i, and $R \cong M_k(S)$. It is now straightforward to show that S has the desired properties. □

We note that 10.6 applies to right Noetherian local rings of finite right global dimension (10.3). Walker proved 10.8 in the case R/J a division ring.

<u>Corollary 10.7</u>. (Ramras) Let R be a right Noetherian local ring of finite right global dimension and assume that R has a right Artinian right quotient ring, then R is a matrix ring over an integral domain.

<u>Proof</u>. By Small's theorem we have C(0) = C(N). But N is torsion with respect to C(N) (10.6). Therefore N = 0 and the result follows from 10.6. □

<u>Corollary 10.8</u>. Let R be a right Noetherian right non-singular local ring of finite right global dimension, then R is a matrix ring over an integral domain.

<u>Proof</u>. Let $x \in N \cap \ell(N)$. Then $xc = 0$ for some $c \in C(N)$ (10.6). Also $xN = 0$ so that $x(cR + N) = 0$. But $(cR + N)/N$ is an essential right ideal of R/N from which it follows easily that $cR + N$ is an essential right ideal of R. Because R is right non-singular we have $x = 0$. Thus $N \cap \ell(N) = 0$.

But $\ell(N)$ is an essential right ideal of R (4.16). Therefore N = 0 and the result follows from 10.6. □

In 10.10 we shall give an example due to J.T. Stafford which shows that a right Noetherian local ring of finite right global dimension need not be semi-prime. That example is in turn based on an example of A.V. Jategaonkar which we shall give in 10.9. What follows is a special case of Jategaonkar's general construction which is of great interest in its own right as being amongst other things a counter-example to the Jacobson conjecture for right Noetherian rings.

Example 10.9. (Jategaonkar) We shall construct an integral domain R in which every right ideal is principal and in which the right ideals are two-sided ideals and are linearly ordered. Let F be a field and let K be the field of rational functions over F in a countable number of indeterminates $x_1, x_2, \ldots$. Let x be another indeterminate and set S = K[x]. Let $p:S \to K$ be the F-algebra homomorphism determined by $p(x_i) = x_{i+1}$ for all i and $p(x) = x_1$, then p maps the non-zero elements of S to units of S.

Let y be another indeterminate and set T = S[y,p], i.e. T is the integral domain whose elements are polynomials in y with coefficients in S written on the right of the powers of y and with addition as usual and multiplication extending the rule sy = yp(s) for all $s \in S$. The usual proof that the ring of polynomials in one indeterminate over a field is a principal ideal domain can be modified to show that every right ideal of T is principal. The key step is to make the leading terms of two polynomials in y match by multiplying one of them on the right by an element of T and this can always be done. For example, if a, $b \in S$ with $b \neq 0$ we can match yb with $y^3a$ by writing $y^3a = yb.y^2a(p(b))^{-2}$ because by = yp(b) and p(b) is a unit of S.

Let C be the set of elements of T which have non-zero constant term when regarded as polynomials in both x and y. We shall show that T satisfies the right Ore condition with respect to C. Let $a \in T$ and $c \in C$. We have $aT + cT = hT$ for some $h \in T$. Thus $a = ha'$, $c = hc'$, and $h = au + cv$ for some $a', c', u, v \in T$. We have $1 = a'u + c'v$. Hence $a'(1 - ua') = c'va'$ and $a'uc' = c'(1 - vc')$. Because $c = hc'$ we have $c' \in C$. If $uc' \notin C$ then u has zero constant term so that $1 - ua' \in C$. Taking $d = uc'$ or $d = 1 - ua'$ we have $a'd = c'b$ for some $d \in C$ and $b \in T$. Therefore $ad = cb$.

Let R be the partial right quotient ring of T with respect to C (the construction of R from T is the same as that of the right quotient ring described in Chapter 1 except that only elements of C are inverted). Let I be a non-zero right ideal of R then $I = (I \cap T)R$. Because every right ideal of T is principal it follows that $I = tR$ for some non-zero element t of T. We have $t = \Sigma\, y^i x^j k_{ij}$ for some $k_{ij} \in K$. Let i, j, m, n be non-negative integers with $i \geq m$ and $j \geq n$ then $y^i x^j = y^m x^n y^{i-m} x^j (p(x))^{-n}$ so that $y^i x^j \in y^m x^n T$. It is now straightforward to show that every non-zero element of T can be written in the form $y^i x^j c$ for some $c \in C$ and hence every non-zero right ideal of R is of the form $y^i x^j R$ (because $cR = R$). Also $y = x^n y(p(x))^{-n}$ for each non-negative integer n from which it follows that $yR = \bigcap_{n=1}^{\infty} x^n R$. Therefore every right ideal of R belongs to the chain

$$0 = \bigcap_{n=1}^{\infty} y^n R \subseteq \ldots \subseteq y^2 R = \bigcap_{n=1}^{\infty} yx^n R \subseteq \ldots \subseteq yxR \subseteq yR = \bigcap_{n=1}^{\infty} x^n R \subseteq \ldots \subseteq xR \subseteq R.$$

It is easy to show that each of these right ideals is two-sided. Clearly xR is the Jacobson radical of R and $(xR)^n = x^n R$ so that yR is the intersection of the powers of the Jacobson radical of R. □

We are very grateful to J.T. Stafford for allowing us to include the following unpublished example.

Example 10.10. (Stafford) With R, x and y as in 10.9 we consider the ring R/yxR which by an abuse of notation we will again call R. Thus $0 \subseteq yR = \bigcap_{n=1}^{\infty} x^nR \subseteq \ldots \subseteq x^2R \subseteq xR \subseteq R$ is a complete list of the right ideals of R, with yx = 0 and xy = ya for some central unit a of R. Clearly R is a right Noetherian local ring which is not semi-prime (because $(yR)^2 = 0$). We shall now show that R has finite right global dimension, in fact rgd(R) = 2. Because xy ≠ 0 it is easy to show that x is right regular. Hence $x^nR$ is a free right R-module for each positive integer n. We have $r(y) = xR$, so that left multiplication by y induces an isomorphism between R/xR and yR. Because xR is not a direct summand of R it follows that yR is not projective, but R and xR are both projective. Thus yR has projective dimension 1 and all other right ideals of R have dimension 0. It now follows from a standard result of homological algebra that rgd(R) = 2. It is straightforward to show that $\ell(y) = yR = Ry$ and hence that the left global dimension of R is infinite. □

Remarks

(1) Let R be a right Noetherian local ring of finite right global dimension then $R/N \cong M_n(D)$ for some integral domain D and positive integer d. Also $R/J \cong M_k(K)$ for some division ring K and positive integer k. Because R/J is isomorphic to a factor ring of R/N it is easy to show that d divides k. It is possible to have d ≠ k. In Section IV of [105] there is a Noetherian integral domain S with a prime ideal P such that P is generated by a central element of S and S/P has Goldie dimension 2. Let R be the localisation of S at P then R is a hereditary Noetherian local domain and R/J is isomorphic to the quotient ring of S/P and so is a 2 by 2 matrix ring over a division ring. Thus d = 1 and k = 2.

(2) Theorem 10.6 appears in [9].

(3) One of the interesting features of Jategaonkar's general construction (of which 10.9 is only a special case) is that it gives a right Noetherian ring with an infinite descending chain of prime ideals.

# 11 The Artin–Rees property

This chapter consists of some results concerning the AR-property (AR being an abbreviation for Artin-Rees). This material is of interest in its own right, but we are giving it now as a preparation for Chapter 12 where we will return to the study of rings with finite global dimension.

An ideal I of a ring R is said to have the *right* AR-*property* if for each right ideal K of R there is a positive integer n such that $K \cap I^n \subseteq KI$. A ring R is said to have the right AR-property, or to be a *right* AR-*ring*, if every ideal of R has the right AR-property. The left AR-property is defined similarly (with K an arbitrary left ideal and $K \cap I^n \subseteq IK$), and if R is both a right AR-ring and a left AR-ring then we call R an AR-ring, etc.

We have already shown that if I is an invertible ideal of a right Noetherian ring then I has the right AR-property (3.3). It is well-known that every commutative Noetherian ring is an AR-ring (this will follow from 11.7), but the following example shows that not every Noetherian ring is an AR-ring.

Example 11.1. Let R be the ring of 2 by 2 upper triangular matrices over a field. Set $I = e_{11}R$ and $K = Re_{22}$ then I and K are ideals of R with $KI = 0$. We have $I = I^2$ so that $K \cap I^n = K \cap I$ for all positive integers n. But $K \cap I \neq 0$, so that I does not have the right AR-property. For later reference we note that I can be generated as a right ideal by $e_{11}$ and $e_{12}$ where $e_{12}$ is a normalising element of R and $e_{11}$ is central $\mathrm{mod}(e_{12}R)$. □

We shall only be interested in the AR-property for rings which are at least right Noetherian. This is why we will usually assume in this chapter that the rings concerned are right Noetherian even though this may not always be necessary.

Lemma 11.2. Let R be a right Noetherian ring and let I be an ideal of R, then the following conditions are equivalent:

(a) for each finitely-generated right R-module M and each submodule K of M there is a positive integer n such that $K \cap MI^n \subseteq KI$;

(b) for each finitely-generated right R-module M and each essential submodule K of M such that $KI = 0$ there is a positive integer n such that $MI^n = 0$.

Proof. It is trivial to show that (a) implies (b). Suppose that (b) is true and let K be any submodule of a finitely-generated right R-module M. Let $\mathcal{S}$ be the set of all submodules C of M such that $K \cap C = KI$ then KI belongs to $\mathcal{S}$. Let C be a maximal element of $\mathcal{S}$ and let D be a submodule of M such that $(K + C) \cap D = C$. We have $C \subseteq D$ and $K \cap D \subseteq (K + C) \cap D = C$. Hence $K \cap D = K \cap C$ so that $K \cap D = KI$. By maximality of C we have $D = C$. Therefore $(K + C)/C$ is an essential submodule of $M/C$. But $KI \subseteq C$ so that $((K + C)/C)I = 0$. Therefore, by (b), there is a positive integer n such that $(M/C)I^n = 0$. Thus $MI^n \subseteq C$ so that $K \cap MI^n \subseteq KI$. □

Corollary 11.3. Let R be a left and right Noetherian right fully bounded ring with Jacobson radical J. Assume that R/J is Artinian, then J has the right AR-property.

Proof. Let M be a finitely-generated right R-module with an essential submodule K such that $KJ = 0$, then K is an Artinian module. Therefore M is Artinian (7.10) so that $MJ^n = 0$ for some n. The result now follows from 11.2. □

Lemma 11.4. (Hartley) Let R be a right Noetherian ring and let I be an ideal of R which has the right AR-property. Let M be a finitely-generated right R-module and let K be a submodule of M, then there is a positive integer n such that $K \cap MI^n \subseteq KI$ (i.e. 11.2(a) is satisfied).

Proof. We shall prove that result by showing that 11.2(b) is satisfied. Thus we assume that M is a finitely-generated right R-module with an essential submodule K such that KI = 0. We shall show that $MI^n = 0$ for some n. Because M is the sum of a finite number of cyclic modules, and because $K \cap xR$ is essential in xR for each cyclic submodule xR of M, we may without loss of generality assume that M is cyclic. Thus there is a surjective homomorphism $f:R \to M$. Set $L = f^{-1}(K)$. Because KI = 0 we have $LI \subseteq \text{Ker}(f)$. But $L \cap I^n \subseteq LI$ for some n. Hence $f(L \cap I^n) = 0$. We have $f(I^n) = MI^n$. Let $x \in I^n$ with $f(x) \in K \cap MI^n$ then $f(x) \in K$ so that $x \in L \cap I^n$. Hence f(x) = 0 so that $K \cap MI^n = 0$. Therefore $MI^n = 0$. □

Corollary 11.5. Let R be a right Noetherian ring with ideals I and A such that I has the right AR-property, then (I + A)/A has the right AR-property as an ideal of R/A.

Proof. Let M be a finitely-generated right R/A-module then we can regard M as being a right R-module with MA = 0. The result now follows easily from 11.4. □

Corollary 11.6. Let R be a right Noetherian ring with an ideal I which has the right AR-property, then $M_n(I)$ has the right AR-property as an ideal of $M_n(R)$.

Proof. For convenience we will write $I_n$ rather than $M_n(I)$, etc. Let K be a right ideal of $R_n$ then we can in a natural way regard K as being a submodule of the finitely-generated right R-module $R_n$. Therefore $K \cap (R_n)I^s \subseteq KI$ for some s (11.4). But $K = KR_n$ and $R_nI = I_n$. Therefore $K \cap (I_n)^s \subseteq KI_n$. □

An ideal I of a ring R is said to have a *centralising set of generators* if there is a finite sequence of elements $x_1,\dots,x_n$ such that
(i) $I = x_1R + \dots + x_nR$, (ii) $x_1$ is a central element of R, and (iii) $x_{i+1}$ is central mod$(x_1R + \dots + x_iR)$ for all i (for example, $x_3 + x_1R + x_2R$ is a central element of $R/(x_1R + x_2R)$). Clearly every finitely-generated ideal of a commutative ring has a centralising set of generators.

Theorem 11.7. (Nouazé and Gabriel) Let R be a right Noetherian ring and let I be an ideal of R which has a centralising set of generators, then I has the right AR-property.

Proof. Let M be a finitely-generated right R-module with an essential submodule K such that KI = 0. We shall show that $MI^n = 0$ for some n and use 11.2. Suppose firstly that I = xR for some central element x of R. Define $f:M \to M$ by $f(m) = mx$ for all $m \in M$, then f is a module homomorphism. By Fitting's lemma there is a positive integer s such that $f^s(M) \cap \mathrm{Ker}(f^s) = 0$. But $K \subseteq \mathrm{Ker}(f)$ so that $K \cap f^s(M) = 0$. Therefore $f^s(M) = 0$, i.e. $0=Mx^s = MI^s$.

Now we turn to the general case where $I = x_1R + \dots + x_kR$ with $x_1$ central in R and $x_{i+1}$ central mod$(x_1R + \dots + x_iR)$. We shall proceed by induction on k. The case k = 1 was dealt with in the last paragraph, so we suppose that $k \geq 2$. From above we also know that $Mx_1{}^t = 0$ for some t. For a fixed value of k we shall proceed by induction on t. Set $H = I/x_1R$. Suppose firstly that t = 1, i.e. that $Mx_1 = 0$, then M is a right $R/x_1R$-module. We have KH = 0 and H has a centralising set of generators with k - 1 elements. Therefore by induction on k there is a positive integer n such that $MH^n = 0$, i.e. $MI^n = 0$.

Now suppose that $t \geq 2$. Set $L = \{m \in M: mx_1 = 0\}$. Because $Lx_1 = 0$ and $K \subseteq L$ is essential in L, as in the last paragraph we have $LI^a = 0$ for some a. Also $Mx_1.x_1^{t-1} = 0$ and $K \cap Mx_1$ is essential in $Mx_1$. Therefore by induction on t we have $Mx_1I^b = 0$ for some b. Thus $MI^bx_1 = 0$ so that $MI^b \subseteq L$. Therefore $MI^{a+b} = 0$. □

In the statement of 11.7 it is not possible to replace "centralising" by "normalising" (11.1).

Corollary 11.8. A commutative Noetherian ring is an AR-ring.

Lemma 11.9. (Smith) Let R be a ring with an ideal T such that T has the right AR-property. Suppose that for each positive integer n the ring $R/T^n$ satisfies the right Ore condition with respect to $C(T/T^n)$, then R satisfies the right Ore condition with respect to $C(T)$.

Proof. Let $a \in R$ and $c \in C(T)$. There is a positive integer n such that $(aR + cR) \cap T^n \subseteq (aR + cR)T = aT + cT$. For such n we know that $R/T^n$ has the right Ore condition with respect to $C(T/T^n)$. Hence $ad - cb \in T^n$ for some $b \in R$ and $d \in C(T)$. Thus $ad - cb \in (aR + cR) \cap T^n$ so that $ad - cb = at + ct'$ for some $t, t' \in T$. Hence $a(d - t) = c(b + t')$ with $d - t \in C(T)$. □

Lemma 11.10. Let R be a left and right Noetherian AR-ring and let I be an ideal of R, then I is torsion as a right R-module with respect to $C(N)$ if and only if I is torsion as a left R-module with respect to $C(N)$.

Proof. Suppose that I is torsion as a right R-module with respect to $C(N)$. We have $I = Rx_1 + \ldots + Rx_k$ for some $x_i \in I$. There exists $c \in C(N)$ such that $x_ic = 0$ for all i (2.2(c)). Thus $Ic = 0$. Set $W = r(I)$ then $c \in W$. We have $I \cap W^n \subseteq IW$ for some n so that $I \cap W^n = 0$. Therefore $W^nI = 0$ so that $c^nI = 0$ with $c^n \in C(N)$. □

Theorem 11.11. Let R be a left and right Noetherian AR-ring then R satisfies the Ore condition with respect to $C(N)$.

Proof. Let k be the smallest positive integer such that $N^k = 0$. We shall proceed by induction on k. If $k = 1$ then $N = 0$ and the result follows by Goldie's theorem. Suppose now that $k \geq 2$. We know that $R/N^{k-1}$ is an AR-ring (11.5), so by induction we know that R has the left and right Ore conditions with respect to $C(N)$ mod$(N^{k-1})$. Set $S = \{x \in R: cx = 0 \text{ for some } c \in C(N)\}$ then S is an ideal of R. For example, let $x \in S$ and $a \in R$. We have $cx = 0$ for some $c \in C(N)$ and so $x \in N$. But $da - bc \in N^{k-1}$ for some $d \in C(N)$ and $b \in R$. Also $N^{k-1}x = 0$. Therefore $(da - bc)x = 0$ so that $dax = bcx = 0$. Hence $ax \in S$.

Let $\rho(M)$ denote the rank of a finitely-generated right R-module M (cf. Chapter 2). Clearly S is torsion as a left R-module with respect to $C(N)$. Hence $\rho(S) = 0$ (11.10). Let $c \in C(N)$ then left multiplication by c induces an isomorphism between $R/r(c)$ and $cR$. Hence $\rho(cR) = \rho(R) - \rho(r(c))$. But $r(c) \subseteq S$ so that $\rho(r(c)) = 0$. Therefore $\rho(R) = \rho(cR)$ so that $\rho(R/cR) = 0$. Thus $R/cR$ is torsion with respect to $C(N)$, so that if $a \in R$ then $ad \in cR$ for some $d \in C(N)$. □

Corollary 11.12. (Smith) Let R be a left and right Noetherian AR-ring and let P be a semi-prime ideal of R, then R satisfies the Ore condition with respect to $C(P)$.

Proof. For each n we know that $P/P^n$ is the nilpotent radical of $R/P^n$, so that $R/P^n$ has the Ore condition with respect to $C(P/P^n)$ (11.11). Therefore R has the Ore condition with respect to $C(P)$ (11.9). □

Example 11.13. As in 5.11 let R be the ring of all matrices of the form

$$\begin{bmatrix} f(0) & g(x) \\ 0 & f(x) \end{bmatrix}$$

then R is right Noetherian. Let P be the nilpotent radical of R then P is prime and P trivially has the AR-property. The elements of $C(P)$ are left regular in R but are not all right regular, so that R does not satisfy the right Ore condition with respect to $C(P)$ (1.30(a)). More directly, set $a = e_{12}$ and $c = xe_{22}$ then $c \in C(P)$, but it is not possible to have $ad = cb$ with $b \in R$ and $d \in C(P)$. □

Remarks

(1) The integral group ring of a finitely-generated nilpotent group is an AR-ring [113] and so also is the universal enveloping algebra of a finite-dimensional nilpotent Lie algebra [104]. The group algebra of a polycyclic-by-finite group G over a field of positive characteristic p is an AR-ring if and only if G is p-nilpotent [120].

(2) For the connection between the AR-property and primary decomposition the reader is referred to [16] and the relevant papers by P.F. Smith.

(3) Let P be a prime ideal of a ring R and suppose that P has the right AR-property. Does R have the right Ore condition with respect to $C(P)$? The answer to this question is "No" if R is only assumed to be right Noetherian (11.13), but the answer is not known if R is left and right Noetherian or if R is a prime right Noetherian ring and $P \neq 0$. It was shown by K.A. Brown, T.H. Lenagan and J.T. Stafford that if R is right Noetherian and P is a prime ideal of R with the right AR-property then R has the right Ore condition with respect to $C(P)$ if and only if P is weakly ideal invariant [7]. No example is known of a left and right Noetherian ring with an ideal which is not

weakly ideal invariant.

(4) Let R be a left and right Noetherian ring with Jacobson radical J. It is not known whether J always has the AR-property in these circumstances.

# 12 AR–rings with finite global dimension

We return to the study of rings with finite global dimension in the special case of AR-rings. The class of Noetherian AR-rings of finite global dimension includes all commutative Noetherian rings of finite global dimension together with certain group rings and their localisations (cf. Chapter 11 Remark (1)). The advantage of the AR-property in this context is that it enables us to use the technique of localisation (11.12). We shall prove for example that a Noetherian AR-ring of finite global dimension is a direct sum of prime rings. We shall also obtain a generalisation of the fact that a commutative Noetherian integral domain of finite global dimension is the intersection of the localisations at its rank 1 primes and that these localisations are local principal ideal domains. Much of the material of this chapter is taken from a paper by K.A. Brown, C.R. Hajarnavis and A.B. MacEacharn. For this chapter only we shall adopt a more sophisticated approach, and the reader who does not wish to get involved in homological algebra and localisation should turn to Chapter 13. We shall state results under convenient assumptions rather than strive for the greatest possible generality.

We shall use N and J to denote the nilpotent and Jacobson radicals respectively of a ring R, and pd(M) will denote the projective dimension of a module M. The following result is well-known and we state it for ease of reference.

Proposition 12.1. Let $0 \to A \to B \to C \to 0$ be an exact sequence of R-modules. Set $a = pd(A)$, $b = pd(B)$, and $c = pd(C)$ then:

(i) $a \le \mathrm{Max}(b, c - 1)$;

(ii) $b \le \mathrm{Max}(a,c)$;

(iii) $c \le \mathrm{Max}(a + 1, b)$;

(iv) if $b < \mathrm{Max}(a,c)$ then $c = a + 1$.

Theorem 12.2. (Bhatwadekar, Goodearl) Let R be a left and right Noetherian ring of finite global dimension n, then there is a simple right R-module S such that $\mathrm{pd}(S) = n$.

Proof. The case $n = 0$ is trivial so we assume that n is positive. We have $n = 1 + \sup(\mathrm{pd}(I))$ where I ranges over the right ideals of R, so that $\mathrm{pd}(I) = n - 1$ for some right ideal I of R. We can only have $\mathrm{pd}(R) = n - 1$ if $n = 1$, in which case R is hereditary and $\mathrm{pd}(I) = n - 1$ for every non-zero right ideal I of R. Therefore there is a right ideal L of R which is maximal with respect to the properties that $L \neq R$ and $\mathrm{pd}(L) = n - 1$.

The next step is to show that R/L is subdirectly irreducible, i.e. that the intersection of all the non-zero submodules of R/L is non-zero. Let $\{L_j\}$ be the set of all right ideals of R which strictly contain L. Set $K = R/L$ and $K_j = R/L_j$ for each j. We have $\mathrm{pd}(L_j) \le n - 2$ for all j so that $\mathrm{pd}(K_j) \le n - 1$ (12.1(iii)). Let wd(M) denote the weak dimension of a module M then, because R is left and right Noetherian, we have $\mathrm{pd}(M) = \mathrm{wd}(M)$ for each finitely-generated right R-module M. Thus $\mathrm{wd}(K_j) \le n - 1$ for all j. Set $I = \bigcap_j L_j$ and $M = \prod_j K_j$, then R/I can be embedded in M. Thus we have an exact sequence $0 \to (R/I) \to M \to M' \to 0$ for some module M'. We have $\mathrm{wd}(M) = \sup(\mathrm{wd}(K_j)) \le n - 1$ and $\mathrm{wd}(M') \le \mathrm{pd}(M') \le n$. By the analogue for weak dimension of 12.1(iv) we cannot have $\mathrm{wd}(R/I) = n$. Hence $\mathrm{pd}(R/I) = \mathrm{wd}(R/I) \le n - 1$ so that $\mathrm{pd}(I) \le n - 2$. Thus $L \subseteq I$ with $\mathrm{pd}(I) \le n - 2$ so that $L \neq I$. Therefore I/L is a simple module.

Set $S = I/L$ then $S$ is simple, and we have an exact sequence $0 \to L \to I \to S \to 0$ with $pd(L) = n - 1$ and $pd(I) \leq n - 2$. Therefore $pd(S) = n$ (12.1(iv)). □

Corollary 12.3. Let R be a left and right Noetherian local ring of finite global dimension n, then $pd(J) = n - 1$.

Proof. This follows immediately from 12.2 and the fact that R/J is the direct sum of a finite number of copies of the unique simple right R-module.□

The first example of a right Noetherian local ring R of finite right global dimension n with $pd(J) \neq n - 1$ was constructed by K.L. Fields, but 10.10 serves the same purpose because in that example we have $rgd(R) = 2$ and $pd(J) = 0$. However, it was shown by M. Boratyński that if R is a right Noetherian local ring of finite right global dimension n and if J satisfies the right AR-property then $pd(J) = n - 1$.

Corollary 12.4. Let R be a left and right Noetherian local ring of finite global dimension n and suppose that the right socle E of R is non-zero, then R is simple Artinian.

Proof. Let S be a simple right R-submodule of E then every simple right R-module is isomorphic to S. Hence $pd(S) = n$ (12.2). But S is a right ideal of R. Therefore $n = 0$. Because R is local it follows that R is simple Artinian. □

Theorem 12.5. (Brown, Hajarnavis and MacEacharn) Let R be a left and right Noetherian AR-ring of finite global dimension then R is semi-prime.

Proof. Suppose that $N \neq 0$ and set $X = r(N)$. Thus $X \neq R$. Let P be an ideal of R which is maximal with respect to the properties that $P \neq R$,

$X \subseteq P$, and P is a right annihilator, then P is prime. Set $Y = \ell(P)$. Suppose that Y is torsion as a right R-module with respect to $C(P)$. Because Y is finitely-generated as a left ideal we have $Yc = 0$ for some $c \in C(P)$, so that $c \in r(Y)$. Thus $c \in P$, which is a contradiction. Let $R_P$ be the localisation of R at P (11.12) then $R_P$ is a left and right Noetherian local ring of finite global dimension. Also $PR_P$ is the Jacobson radical of $R_P$ and $YPR_P = 0$. Because Y is not torsion with respect to $C(P)$ we know that $YR_P \neq 0$. Thus $PR_P$ has non-zero left annihilator in $R_P$, so that the right socle of $R_P$ is non-zero. Therefore $R_P$ is a simple Artinian ring (12.4), so that $PR_P = 0$. Thus $cP = 0$ for some $c \in C(P)$ so that $cX = 0$. But $d \in X$ for some $d \in C(N)$ (10.6 and 10.3). Thus $cd = 0$ so that $c \in N$. Therefore $c \in P$, which is a contradiction. □

Corollary 12.6. Let R be a left and right Noetherian local AR-ring of finite global dimension then R is prime.

Proof. We have $N = 0$ (12.5). Now use 10.6 and 10.3. □

Corollary 12.7. Let R be a left and right Noetherian AR-ring of finite global dimension then R is a direct sum of prime rings.

Proof. By 12.5 we know that R is semi-prime. Let M be a maximal ideal of R. There is a one-to-one correspondence between the prime ideals of R which are contained in M and the prime ideals of the localisation $R_M$ of R at M. But $R_M$ is a prime ring (12.6). Therefore M contains only one minimal prime ideal of R. Thus if P and Q are distinct minimal primes of R then $P + Q = R$. It now follows by a standard argument that R is isomorphic to the direct sum of the rings $R/P$ where P ranges over the minimal primes of R. □

We shall not go any further in the general study of Noetherian AR-rings of finite global dimension. We shall turn now to the special case where the rings have Krull dimension 1, and apart from being of interest in its own right this material will be used to derive an intersection theorem for certain prime Noetherian AR-rings of finite global dimension.

Theorem 12.8. (Brown, Hajarnavis, MacEacharn) Let R be a left and right Noetherian local ring of finite global dimension and assume that R has Krull dimension 1, then R is a hereditary prime ring in which every one-sided ideal is principal and every non-zero two-sided ideal is a power of J.

Proof. We have $Nc = 0$ for some $c \in \mathcal{C}(N)$ (10.6 and 10.3). Set $I = N \cap \ell(N)$ then $I(cR + N) = 0$. But $R/(cR + N)$ is an Artinian module because R/N has the right restricted minimum condition. It follows easily that I is Artinian. Because R has Krull dimension 1 we know that R is not Artinian. Therefore the right socle of R is zero (12.4) so that $I = 0$. Hence $N = 0$ because $\ell(N)$ is an essential right ideal of R. Therefore R is a prime ring (10.6 and 10.3).

Let c be a regular element of J then R/cR is a non-zero Artinian module. Let S be the unique simple right R-module and set $\text{rgd}(R) = n$. We have $\text{pd}(S) = n$ (12.2). Also R/cR has a composition series whose factors are isomorphic to S. It follows easily that $\text{pd}(R/cR) = n$. But $\text{pd}(cR) = 0$ because c is regular. Hence $\text{pd}(R/cR) = 1$ so that $n = 1$.

Let P be the unique finitely-generated indecomposable projective right R-module (10.3) and let E be an essential right ideal of R. Because E is projective we have $E \cong P^i$ for some positive integer i. Also $R \cong P^j$ for some j. But E and R have the same Goldie dimension, so that $i = j$. Thus $E \cong R$ so that E is free of rank 1 as a right R-module. Therefore $E = dR$

for some regular element d of R. Because every right ideal is a direct summand of an essential right ideal it follows that every right ideal of R is principal.

In particular we have $J = cR = Rd$ for some regular elements c and d of R. Set $X = Rc^{-1}$ and $Y = d^{-1}R$ then $R = XJ = JY$. Hence $X = XR = XJY = RY = Y$. Thus J is invertible with $J^{-1} = X$. Let I be a non-zero ideal of R with $I \subseteq J$ then $XI \subseteq R$. Let k be the largest positive integer such that $X^kI \subseteq R$ then $X^kI$ is not contained in J. But $X^kI$ is an ideal of R. Therefore $X^kI = R$ so that $I = X^{-k} = J^k$. □

The methods used in the proof of 12.8 can also be used to prove the next two results.

Proposition 12.9. Let R be a left and right Noetherian local ring of global dimension at most 1, then either R is simple Artinian or R is a prime hereditary ring as in 12.8.

Proposition 12.10. Let R be a left and right Noetherian local ring of finite global dimension and assume that R has a non-zero projective right ideal I such that R/I is a non-zero Artinian module, then either R is simple Artinian or R is a prime hereditary ring as in 12.8.

Proof. We have $pd(R/I) = 0$ or 1 according as I is or is not a direct summand of R. As in the proof of 12.8 it follows that the unique simple right R-module S has $pd(S) = 0$ or 1, etc. □

Corollary 12.11. Let R be a left and right Noetherian local ring of finite global dimension and suppose that there is an invertible ideal X of R such that J is minimal amongst prime ideals which contain X, then R is a prime hereditary ring as in 12.8.

Proof. Clearly $J/X$ is the nilpotent radical of $R/X$ so that $R/X$ is Artinian. Because $X$ is invertible it is easy to show that $X$ is projective and the result follows from 12.10. □

In the intersection theorem which we are aiming for we shall assume that every non-zero ideal of the ring concerned contains an invertible ideal. This is why we consider invertible ideals in 12.11 and the next two results.

Lemma 12.12. (Brown, Hajarnavis, MacEacharn) Let $R$ be a left and right Noetherian AR-ring of finite global dimension and let $X$ be an invertible ideal of $R$ with $X \neq R$. Let $P$ be a prime ideal of $R$ such that $X \subseteq P$ and $P/X$ is a right annihilator in $R/X$, then $P/X$ is a minimal prime of $R/X$.

Proof. Let $L$ be the ideal of $R$ such that $X \subseteq L$ and $L/X$ is the left annihilator of $P/X$ in $R/X$. Because $P/X$ is a right annihilator we know that if $r \in R$ with $Lr \subseteq X$ then $r \in P$. Let $T$ be the localisation of $R$ at $P$, then $T$ is a left and right Noetherian local ring of finite global dimension, say $\mathrm{rgd}(T) = t$. For every ideal $I$ of $R$ we have $IT = TI$, and $PT$ is the Jacobson radical of $T$. If $LT = XT$ then $Lc \subseteq X$ for some $c \in C(P)$, which is not so. Therefore $XT$ is strictly contained in $LT$. We have $LT.PT \subseteq XT$ so that $LT/XT$ is a non-zero Artinian right $T$-module. Let $K$ be a right ideal of $T$ such that $XT \subseteq K$ and $K/XT$ is a simple submodule of $LT/XT$, then $\mathrm{pd}(K/XT) = t$ (12.2). But $X$ is an invertible ideal of $R$ so that $XT$ is an invertible ideal of $T$. Therefore $XT$ is projective. Because $T$ is not semi-simple Artinian we cannot have $\mathrm{pd}(K) = t$. Hence $t = 1$ (12.1(iv)). Thus $XT = (PT)^i$ for some positive integer $i$ (12.9). Therefore $PT/XT$ is a minimal prime of $T/XT$ so that $P/X$ is a minimal prime of $R/X$. □

Corollary 12.13. Let $R$ be a left and right Noetherian AR-ring of finite global dimension and let $X$ be an invertible ideal of $R$ with $X \neq R$, then $R/X$

has an Artinian quotient ring.

Proof. Let K be the nilpotent radical of R/X. Set $B = \{y \in R/X: yc = 0$ for some $c \in C(K)\}$. We know that R/X satisfies the right Ore condition with respect to $C(K)$ (11.11 and 11.5), so that B is an ideal of R/X. Because R/X is left Noetherian we have $Bc = 0$ for some $c \in C(K)$. No minimal prime ideal of R/X contains c (1.25), so that no right annihilator prime of R/X contains $r(B)$ (12.12). Therefore $r(B) = R/X$ so that $B = 0$. Hence R/X has an Artinian quotient ring, by Small's theorem. □

Notation 12.14. Let R be a prime left and right Noetherian ring with quotient ring Q and assume that every non-zero ideal of R contains an invertible ideal of R. If $q \in Q$ and X is an invertible ideal of R then $qX \subseteq R$ if and only if $q \in X^{-1}$, so that $qX \subseteq R$ if and only if $Xq \subseteq R$. Set $S = \{q \in Q: qX \subseteq R$ for some invertible ideal X of R$\}$, then S is a subring of Q and $R \subseteq S$. Let $q \in S$ with $qX \subseteq R$ where X is an invertible of R. We have $q \in qXX^{-1}$ with $qX \subseteq R$ and $X^{-1} \subseteq S$ so that $q \in (qS \cap R)S$. Therefore for each right ideal I of S we have $I = (I \cap R)S$ from which it follows that S is right Noetherian. Similarly S is left Noetherian. Let I be a non-zero ideal of S then $I \cap R$ is a non-zero ideal of R. Hence $I \cap R$ contains an invertible ideal X of R so that I contains $XX^{-1}$, i.e. $I = S$. Thus S is a simple ring. It can be shown that if R has finite global dimension then so also does S.

Theorem 12.15. (Brown, Hajarnavis, MacEacharn) Let R be a left and right Noetherian prime AR-ring of finite global dimension and assume that every non-zero ideal of R contains an invertible ideal of R. Let S be as in 12.14, then $R = S \cap (\cap R_P)$ where P ranges over the rank 1 primes of R and $R_P$ denotes the localisation of R at P. Also if P is a rank 1 prime of R then $R_P$ is a left and right Noetherian prime hereditary local ring in which every one-sided ideal is principal and every non-zero ideal is a power of the Jacobson radical of $R_P$.

Proof. Let P be a rank 1 prime of R then P contains an invertible ideal X of R. Clearly P is minimal over X so that $PR_P$ is the only prime ideal of $R_P$ which contains $XR_P$. Also $XR_P$ is an invertible ideal of $R_P$. It now follows from 12.11 that $R_P$ has the desired properties.

Let $s \in S \cap (\cap R_P)$ where P ranges over the set $\mathcal{W}$ of rank 1 primes of R. Set $K = \{r \in R: sr \in R\}$, then K is a right ideal of R. For each $P \in \mathcal{W}$ we know that K contains an element of $C(P)$. Also there is an invertible ideal X of R such that $X \subseteq K$. Let $P_1,\ldots,P_k$ be the prime ideals of R which are minimal over X, then $P_i \in \mathcal{W}$ for all i (3.4). Thus for each i we know that K contains an element of $C(P_i)$. Hence there exists $c \in K$ such that $c \in C(P_i)$ for all i (a more general version of this claim will be proved in Chapter 13). Thus $c \in C(P_1 \cap \ldots \cap P_k)$ so that c + X is regular modulo the nilpotent radical of R/X. But R/X has an Artinian quotient ring (12.13), so that $c \in C(X)$. We have $sc \in R$ so that $Xsc \subseteq X$. But $Xs \subseteq R$ and $c \in C(X)$. Therefore $Xs \subseteq X$ so that $X^{-1}Xs \subseteq X^{-1}X$, i.e. $Rs \subseteq R$. Hence $s \in R$. □

Corollary 12.16. Let R be a left and right Noetherian prime AR-ring of finite global dimension and assume that R satisfies a polynomial identity, then $R = \cap R_P$ where P ranges over the rank 1 primes of R and each $R_P$ is as in 12.15.

Proof. Let I be a non-zero ideal of R then I contains a non-zero central element a of R. Clearly aR is an invertible ideal of R. Thus 12.15 applies to R. Let c be a regular element of R then cR contains a non-zero central element of R. It follows that S = Q in the notation of 12.14. □

Corollary 12.17. Let R be a prime left and right Noetherian ring of finite global dimension and assume that every ideal of R has a centralising set of generators, then $R = S \cap (\cap R_p)$ as in 12.15.

Proof. Clearly each non-zero ideal of R contains a non-zero central element and hence also an invertible ideal. Also R is an AR-ring (11. 7). □

The following example shows that a prime local Noetherian ring R need not have finite global dimension even if the centre C of R has finite global dimension and R is closely related to C in the sense that R is finitely-generated as a C-module and every ideal of R has a centralising set of generators.

Example 12.18. Let S be Z localised at 2Z and set

$$R = \left\{ \begin{bmatrix} a & 2b \\ c & d \end{bmatrix} : a, b, c, d \in S, \ a - d \in 2S \right\}$$

with the usual matrix operations, then it is easy to show that R is a prime ring. The centre C of R is the set of all scalar matrices in R so that $C \cong S$. Because R is contained in the finitely-generated C-module $M_2(S)$ we know that R is finitely-generated as a C-module and hence that R is left and right Noetherian. Set $J = 2Se_{11} + 2Se_{12} + Se_{21} + 2Se_{22}$ then J is an ideal of R with $R/J \cong Z/2Z$. Let $r \in R$ with $r \notin J$ then the diagonal entries of r do not belong to 2S. Hence $\det(r) \notin 2S$ so that $\det(r)$ is a unit of S and r is a unit of R. Therefore J is the Jacobson radical of R. Because R is not an integral domain and R/J is a field it follows from 10.7 or 10.8 that R does not have finite global dimension (cf. Chapter 10 Remark (1)). We can also show this more directly by noting that $r(e_{21}) = e_{21}R$ so that there is an exact sequence $0 \to e_{21}R \to R \to e_{21}R \to 0$. Also $e_{21}R$ is not projective because $r(e_{21})$ is not generated by an idempotent. Therefore $\mathrm{pd}(e_{21}R)$

is not finite (12.1(iv)).

Set $x = e_{21} + 2e_{12}$ then $xR = Rx$. We have $J^2 = 2Se_{11} + 4Se_{12} + 2Se_{21} + 2Se_{22}$ so that $J/J^2$ is a 2-dimensional vector space over $Z/2Z$. Therefore there are three ideals of R strictly between J and $J^2$ and these can be identified as $xR$, $A = Re_{21}R = 2Se_{11} + 4Se_{12} + Se_{21} + 2Se_{22}$, and $B = R2e_{12}R = M_2(2S)$. We have $J = xR \cup A \cup B$ so that J is not principal. It is straightforward to show that $2R \subseteq J^2$ and that $R/2R$ is commutative, from which it follows that J, xR, A and B have centralising sets of generators.

We shall prove that every non-zero ideal of R is of the form $x^nR$, $x^nA$, $x^nB$, or $J^n$ for some non-negative integer n. One consequence of this is that every ideal of R has a centralising set of generators (because the set of ideals with centralising sets of generators is closed under multiplication), and another is that the following is a complete description of the lattice of proper ideals of R: J contains xR, A, and B, each of which contains $J^2$ ; $J^2$ contains $x^2R$, xA, and xB, each of which contains $J^3$; and so on. Let U be a non-zero ideal of R and let k be the largest integer such that $x^{-k}U \subseteq R$. Set $V = x^{-k}U$ then V is not contained in xR. We shall show that V = R or J or A or B. If V is not contained in J then V = R. Suppose now that $V \subseteq J$. We shall deal with two cases.

Case (a): There exists $v \in V$ with $v \notin xR$ and $v \in A$. We have $v = 2ae_{11} + 4be_{12} + ce_{21} + 2de_{22}$ for some $a, b, c, d \in S$ with $c \notin 2S$. Because c is a unit of S we may suppose that c = 1. By considering $(e_{11} - 2ae_{12} + e_{22})v$ we see that without loss of generality we may take a = 0. By considering $v(e_{11} - 2de_{12} + e_{22})$ we may take d = 0. Thus $v = 4be_{12} + e_{21}$. We have $2e_{12}v . 2e_{12} = 4e_{12}$. Hence $4e_{12} \in V$ and $4be_{12} + e_{21} \in V$ so that $e_{21} \in V$. Therefore $A \subseteq V$.

Case (b): There exists $v \subset V$ with $v \notin xR$ and $v \in B$. We have $v = 2ae_{11} +$

$2be_{12} + 2ce_{21} + 2de_{22}$ for some a, b, c, d $\in$ S and b $\notin$ 2S. Because b is a unit of S we can take b = 1. By considering $v(e_{11} - ae_{21} + e_{22})$ we can take a = 0. By considering $(e_{11} - de_{21} + e_{22})v$ we can take d = 0. Thus $v = 2e_{12} + 2ce_{21}$. We have $e_{21}ve_{21} = 2e_{21}$. Hence $2e_{21} \in V$ and $2e_{12} + 2ce_{21} \in V$ so that $2e_{12} \in V$. Therefore $B \subseteq V$.

Thus we have $V \subseteq J$, and $A \subseteq V$ or $B \subseteq V$. Therefore V = J or A or B. □

Remarks

(1) Further examples can be found in [9] where in particular it is shown that: (i) For each positive integer n there is a local Noetherian AR-ring R of global dimension n such that the Jacobson radical of R is the only non-zero prime ideal of R; (ii) There is a local Noetherian ring R of global dimension 3 such that every ideal of R has a centralising set of generators, every saturated chain of prime ideals of R has length 3, there is only one prime ideal of R which has rank 1 (cf. 12.15), there is no non-zero central element of R which belongs to J and not to $J^2$, and R/J is a division ring and is the only proper factor ring of R which has finite global dimension.

(2) For further information concerning rings with finite global dimension the reader is referred to [9] and its proposed second part.

(3) J.C. Robson has given an example of a left and right Noetherian hereditary integral domain R such that R has a unique non-zero ideal I with I ≠ R [118, Example 7.3].

(4) If R is a left and right Noetherian prime local ring such that the Jacobson radical of R is projective, then R is a principal right and left ideal ring [73].

# 13 Noetherian quotient rings

Let R be a left and right Noetherian ring and let A and J be the Artinian and Jacobson radicals respectively of R. The central result of this chapter, which is due to J.T. Stafford, is that R is its own quotient ring (i.e. every regular element of R is a unit) if and only if $\ell(A) \cap r(A) \subseteq J$. It follows easily that if R is its own quotient ring then so also is $M_n(R)$, and also that R/J is Artinian and $A \neq 0$. At the end of the chapter we shall give some examples to show what happens if R is only assumed to be right Noetherian. Much of the material of this chapter is due to J.T. Stafford.

We shall use A(R) and J(R), or just A and J, to denote the Artinian and Jacobson radicals respectively of a ring R. We will usually assume that R is left and right Noetherian and will not strive for generality. The following piece of notation will be used throughout the chapter.

Notation 13.1. Let R be a left and right Noetherian ring and let $P_1$ be a maximal left annihilator prime in R, i.e. $P_1$ is an ideal of R which is maximal with respect to being of the form $\ell(I)$ for some non-zero ideal I of R ($P_1$ is necessarily prime). Set $B_1 = r(P_1)$. Let $P_2$ be a prime ideal of R such that $B_1 \subseteq P_2$ and $P_2/B_1$ is a maximal left annihilator prime in $R/B_1$. Let $B_2$ be the ideal of R such that $B_1 \subseteq B_2$ and $B_2/B_1 = r(P_2/B_1)$ in $R/B_1$. Continuing this process we obtain ideals $0 = B_0 \subsetneqq B_1 \subsetneqq \cdots \subsetneqq B_m = R$ and prime ideals $P_1,\ldots,P_m$ such that $B_{i-1} \subseteq P_i$, and also in $R/B_{i-1}$ we have $P_i/B_{i-1} = \ell(B_i/B_{i-1})$ and $B_i/B_{i-1} = r(P_i/B_{i-1})$. The $B_i$ are said to form a *left affiliated series* for R and the $P_i$ to form a set of *left affiliated primes* for R. A right affiliated series can be defined similarly. The process given above for forming the $B_i$ stops when say $B_{m-1}$ is a prime ideal, and then $P_m = B_{m-1}$ and $B_m = R$.

In future we will refer to the notation of 13.1 without further explanation. The reason for using affiliated series is seen in 13.3 where we shall show that if $c \in C(P_i)$ for all i then c is right regular.

Lemma 13.2. In the notation of 13.1, the left $R/P_i$-module $B_i/B_{i-1}$ is torsion-free.

Proof. It is enough to show that $B_1$ is torsion-free as a left $R/P_1$-module. Set $I = \{x \in B_1 : cx = 0 \text{ for some } c \in C(P_1)\}$ then I is an ideal of R. Because I is finitely-generated as a right ideal we have $cI = 0$ for some $c \in C(P_1)$. Thus $RcR + P_1 \subseteq \ell(I)$. Because $P_1$ is a maximal left annihilator prime we must have $I = 0$. □

Lemma 13.3. With the notation of 13.1 suppose that $c \in C(P_i)$ for all i then c is right regular.

Proof. Let $x \in R$ with $cx = 0$. Thus $x \in B_m$ and $cx \in B_{m-1}$ with $c \in C(P_m)$, so that $x \in B_{m-1}$ (13.2). But $cx \in B_{m-2}$ with $c \in C(P_{m-1})$ so that $x \in B_{m-2}$, and so on. Eventually we have $x \in B_0$, i.e. $x = 0$. □

The next result will be useful in combination with 13.3.

Lemma 13.4. Let R be a right Noetherian ring and let $X_1,\ldots,X_s$ be prime ideals of R. Let K be a right ideal of R and assume for each i that K contains an element of $C(X_i)$, then there exists $c \in K$ such that $c \in C(X_i)$ for all i.

Proof. We may suppose that the $X_i$ are distinct and that they have been arranged so that if $i < j$ then $X_i$ is not contained in $X_j$. Also by induction we may assume that there exists $a \in K$ such that $a \in C(X_1) \cap \ldots \cap C(X_{s-1})$.

Set $Y = X_1X_2\ldots X_{s-1}$ then $(Y + X_s)/X_s$ is a non-zero ideal of $R/X_s$. Also $(K + X_s)/X_s$ is an essential right ideal of $R/X_s$. Let I be a right ideal of R such that $X_s \subseteq I$ and $I \cap KY \subseteq X_s$ then $(I \cap K)Y \subseteq X_s$ so that $I \cap K \subseteq X_s$. Hence $I \cap (K + X_s) = X_s$ so that $I = X_s$. Thus $(KY + X_s)/X_s$ is an essential right ideal of $R/X_s$. Therefore there exists $y \in KY$ such that $a + y \in C(X_s)$ (1.20; in fact it is enough to combine the first paragraph of the proof of 1.18 with 1.17 and 1.13). Set $c = a + y$ then $c \in K \cap C(X_1) \cap \ldots \cap C(X_s)$. □

<u>Lemma 13.5</u>. Let R be a left and right Noetherian ring then there is a left affiliated series for R as in 13.1 with $B_k = A(R)$ for some k.

<u>Proof</u>. We shall write A instead of A(R). If $A = 0$ we can take $B_0 = A$. Suppose now that $A \neq 0$. Let $P_1$ be an ideal of R which is maximal with respect to being of the form $\ell(I)$ where I is a non-zero ideal of R with $I \subseteq A$. We have $r(\ell(A)) = A$ (4.9(d)). Also $\ell(A) \subseteq P_1$ so that $r(P_1) \subseteq A$. Hence if X is an ideal of R with $P_1 \subseteq X$ and $r(X) \neq 0$ then $X = P_1$. Thus $P_1$ is a maximal left annihilator prime. Set $B_1 = r(P_1)$ then $B_1 \subseteq A$. Clearly $A/B_1 = A(R/B_1)$ and the process can be repeated in $R/B_1$. □

<u>Lemma 13.6</u>. Let R be a left and right Noetherian ring with $A(R) = 0$, then in the notation of 13.1 we have $A(R/B_i) = 0$ for all i.

<u>Proof</u>. It is enough to show that $A(R/B_1) = 0$. Let X be the ideal of R such that $B_1 \subseteq X$ and $X/B_1 = A(R/B_1)$. Let $T = \{r \in R: Xr \subseteq B_1\}$, then $T/B_1 = r(X/B_1)$ in $R/B_1$. But R/T is Artinian (4.5(c)). Also $XT \subseteq B_1$ so that $P_1XT = 0$. Hence $P_1X \subseteq A(R)$ so that $P_1X = 0$. Therefore $X \subseteq r(P_1)$. Hence $X = B_1$. □

<u>Corollary 13.7</u>. Let R be a left and right Noetherian ring then there is a left affiliated series for R as in 13.1 with $B_k = A(R)$ for some k. Also

$R/P_i$ is Artinian if and only if $i \le k$, and $\ell(A(R)) \subseteq P_i$ if $i \le k$.

Proof. There is a left affiliated series with $B_k = A(R)$ for some k (13.5). If $i \le k$ then $P_i/B_{i-1}$ is the left annihilator of an ideal contained in $A(R)/B_{i-1}$ so that $\ell(A(R)) \subseteq P_i$, and hence $R/P_i$ is Artinian (4.5(c)). Now suppose that $k \le i$. We have $A(R/B_k) = A(R/A(R)) = 0$ so that $A(R/B_i) = 0$ (13.6). Hence $B_{i+1}/B_i$ is not an Artinian left ideal of $R/B_i$. Therefore $R/P_{i+1}$ is not Artinian. □

Corollary 13.8. Let R be a left and right Noetherian ring with $A(R) = 0$ and assume that R/P is Artinian for each non-minimal prime ideal P of R, then R has an Artinian quotient ring.

Proof. Let $P_1,\ldots,P_m$ be a set of left affiliated primes for R as in 13.1. Because $A(R) = 0$ we know that each $R/P_i$ is not Artinian (13.7). Therefore each $P_i$ is a minimal prime of R. Let $c \in C(N)$ then $c \in C(P)$ for each minimal prime P of R. Therefore $c \in C(P_i)$ for all i, so that c is right regular (13.3). By symmetry c is left regular. Therefore R has an Artinian quotient ring by Small's theorem. □

Corollary 13.9. (Lenagan) Let R be a left and right Noetherian ring of right Krull dimension 1 with $A(R) = 0$, then R has an Artinian quotient ring.

Proof. This follows easily from 13.8 because if P is a prime ideal of R which is not minimal then P/N is an essential right ideal of R/N. □

Theorem 13.10. (Stafford) Let R be a left and right Noetherian ring then R is its own quotient ring if and only if $\ell(A) \cap r(A) \subseteq J$ where A and J are the Artinian and Jacobson radicals respectively of R.

Proof. Suppose that R is its own quotient ring and let M be a maximal right ideal of R. Let $P_1, \ldots, P_m$ be a set of left affiliated primes for R as in 13.7 with $B_k = A$ for some k, and let $Q_1, \ldots, Q_n$ be a similar set of right affiliated primes. We have $C(P_1) \cap \ldots \cap C(P_m) \cap C(Q_1) \cap \ldots \cap C(Q_n) \subseteq C(0)$ (13.3). Because R is its own quotient ring we know that M does not contain a regular element of R. Therefore there exists i such that M does not contain an element of $C(P_i)$ or there exists j such that M does not contain an element of $C(Q_j)$ (13.4). Suppose that M does not contain an element of $C(P_i)$. Hence $R \neq M + P_i$. Therefore $P_i \subseteq M$. Also $M/P_i$ cannot be an essential right ideal of $R/P_i$ by Goldie's theorem. Because $M/P_i$ is a maximal right ideal of $R/P_i$ it follows easily that the socle of $R/P_i$ is non-zero so that $R/P_i$ is Artinian (1.24). Therefore $\ell(A) \subseteq P_i$ (13.7) so that $\ell(A) \subseteq M$. If there exists j such that M does not contain an element of $C(Q_j)$ then similarly we have $r(A) \subseteq M$. Therefore $\ell(A) \cap r(A) \subseteq M$ for every maximal right ideal M of R so that $\ell(A) \cap r(A) \subseteq J$.

Conversely set $T = \ell(A) \cap r(A)$ and assume that $T \subseteq J$. By 4.5(c) we know that $R/\ell(A)$ and $R/r(A)$ are Artinian rings and hence so also is R/T. Let c be a regular element of R. It is easy to show that $c + \ell(A)$ is a right regular element of $R/\ell(A)$ so that $c \in C(\ell(A))$ (13.11). Similarly $c \in C(r(A))$ so that $c \in C(T)$. Therefore c + T is a unit of R/T (13.11). But $T \subseteq J$ so that c + J is a unit of R/J. Hence c is a unit of R. □

Lemma 13.11. Let R be a right Artinian ring and let c be a right regular element of R then c is a unit of R.

Proof. We have $cR \supseteq c^2R \supseteq \ldots$ so that $c^nR = c^{n+1}R$ for some n. Hence $c^n = c^{n+1}d$ for some $d \in R$. Because c is right regular we have 1 = cd so that d is also right regular. Similarly da = 1 for some $a \in R$ from which it follows that cd = dc = 1. □

Corollary 13.12. Let R be a left and right Noetherian ring which is its own quotient ring then $A(R) \neq 0$ and $R/J$ is Artinian.

Proof. By 13.10 we have $\ell(A) \cap r(A) \subseteq J$ so that $A \neq 0$. It was shown in the last paragraph of the proof of 13.10 that $R/J$ is Artinian. □

Corollary 13.13. Let R be any ring which has a left and right Noetherian quotient ring Q, then $M_n(Q)$ is the quotient ring of $M_n(R)$.

Proof. Let I be the identity element of $M_n(R)$. Let $x \in M_n(Q)$ and let $x_{ij}$ denote the (i,j)-entry of x. There is a regular element c of R such that $x_{ij}c \in R$ for all i and j. Set $s = cI$ then s is a regular element of $M_n(R)$ and $xs \in M_n(R)$. It follows easily that a right regular element of $M_n(R)$ is right regular in $M_n(Q)$. By symmetry, every regular element of $M_n(R)$ is regular in $M_n(Q)$. Thus it is enough to prove that $M_n(Q)$ is its own quotient ring. This follows immediately from 13.10 because $J(M_n(Q)) = M_n(J(Q))$ and $A(M_n(Q)) = M_n(A(Q))$. □

Proposition 13.14. Let R be a left and right Noetherian ring with $A(R) = 0$ then every maximal right ideal of R contains a regular element.

Proof. This requires only minor modifications to the first part of the proof of 13.10. □

Corollary 13.15. Let R be a left and right Noetherian ring with $A(R) = 0$ and let M be a right R-module of finite length then M is torsion with respect to the regular elements of R.

Proof. Without loss of generality we may suppose that M is simple. Let x be a non-zero element of M and set $K = \{r \in R: xr = 0\}$ then K is a maximal right ideal of R. Therefore K contains a regular element c of R (13.14) and we have $xc = 0$. □

Let R be a right Noetherian ring which is its own quotient ring then R/J need not be Artinian (13.16) and R need not have any non-zero Artinian right ideals (cf. 13.12). To justify the latter claim we need only consider the right quotient ring of the ring given in 5.11 (i.e. modify 5.11 by letting f(x) and g(x) belong to the localisation of F[x] at the principal ideal generated by x).

Example 13.16. (Stafford) We shall construct a ring R which is right Noetherian and is its own quotient ring but R/J is not Artinian. Let V be a (non-Artinian) right Noetherian simple ring which has (up to isomorphism) a unique simple right V-module S (such a ring has been constructed by J.H. Cozzens). Let K be the endomorphism ring of S then K is a division ring. Set

$$R = \begin{bmatrix} K & S \\ 0 & V \end{bmatrix} .$$

The nilpotent radical N of R consists of the strictly upper triangular elements of R. Therefore N is a principal right ideal of R because S is a cyclic right V-module. Also $R/N \cong K \oplus V$ which is right Noetherian, so that R is right Noetherian. Also N is the Jacobson radical of R and R/N is not Artinian.

Let c be a regular element of R then $c = ke_{11} + se_{12} + ve_{22}$ for some $k \in K$, etc. Clearly $k \neq 0$, so that in order to prove that c is a unit of R it is enough to show that v is a unit of V. Suppose that $vV \subseteq M$ for some maximal right ideal M of V. There is an isomorphism $f: V/M \to S$. Set $t = f(1 + M)$. Because $v \in M$ we have $tv = f(1 + M)v = f(v + M) = 0$. Set $r = te_{12}$ then $r \in R$ with $rc = 0$ and $r \neq 0$, which is a contradiction. Therefore $vV = V$ so that $vw = 1$ for some $w \in V$. Clearly w is right regular in V and hence is regular (1.13). Because $w = wvw$ we also have $wv = 1$. □

Example 13.17. We shall construct a right Noetherian ring R such that R has a right quotient ring but $M_2(R)$ does not (cf. 13.13). The construction is in several stages and we start by taking T to be a right Noetherian integral domain which does not have left Goldie dimension 1 (e.g. 1.22). Let u be an indeterminate which commutes with the elements of T and let C be the set of all elements of T[u] which have non-zero constant term. Because T[u] is a right Noetherian integral domain we know that T[u] satisfies the right Ore condition with respect to its non-zero elements. From this it follows easily that T[u] satisfies the right Ore condition with respect to C. Let S be the partial right quotient ring of T[u] with respect to C and let D be the right quotient division ring of T. We make D into a right T[u]-module by setting Du = 0. Set

$$R = \begin{bmatrix} I & T \\ 0 & T[u] \end{bmatrix} \quad \text{and} \quad Q = \begin{bmatrix} D & D \\ 0 & S \end{bmatrix}$$

An element of R is regular if and only if it has a non-zero element of T in the (1,1)-position and an element of C in the (2,2)-position. It is now not hard to show that Q is the right quotient ring of R.

We shall now show that $M_2(R)$ does not have a right quotient ring. There are non-zero elements p and q of T such that $Tp \cap Tq = 0$. Thus if $s, t \in T$ with $sp = tq$ then $s = t = 0$. Working in R set

$$x = \begin{bmatrix} p & 0 \\ 0 & p \end{bmatrix}, \quad y = \begin{bmatrix} q & 0 \\ 0 & q \end{bmatrix} \quad \text{and} \quad z = \begin{bmatrix} 1 & 0 \\ 0 & u \end{bmatrix}$$

It is easy to show that x and y are regular elements of R and that $Rx \cap Ry = 0$. Thus if $a, b \in R$ with $ax + by = 0$ then $a = b = 0$. Also z is right regular but not left regular in R. Working in $M_2(R)$ set $a = e_{11}$ and $c = ze_{11} + xe_{12} + ye_{22}$

then c is a regular element of $M_2(R)$. Suppose that b, d $\in M_2(R)$ with ad = cb, and let $b_{ij}$ and $d_{ij}$ be the (i,j)-entries of b and d respectively. By comparing the second rows of ad and cb we see that $0 = yb_{21} = yb_{22}$ so that $b_{21} = b_{22} = 0$. By comparing the first rows we have $d_{11} = zb_{11}$ and $d_{12} = zb_{12}$. Because z is not left regular in R it follows that d is not left regular in $M_2(R)$. Hence $M_2(R)$ does not satisfy the right Ore condition with respect to its regular elements. Although c is a regular element of $M_2(R)$ it is only right regular as an element of $M_2(Q)$, and this in a sense is why $M_2(Q)$ is not the right quotient ring of $M_2(R)$ (cf. proof of 13.13). Clearly the elements a and c given above can also be used to show that the ring of 2 by 2 upper triangular matrices over R does not have a right quotient ring. □

Remarks

(1) Example 13.17 is taken from [26] where more detail is given. Also in [26] there is an example of a left and right Noetherian P.I. ring R which is its own quotient ring and which has an idempotent element e such that eRe has neither a right nor a left quotient ring.

(2) For further details concerning regular elements and quotient rings see for example [130], [150], and [152].

(3) Let R be a right Noetherian ring. The following questions are open: If $M_2(R)$ has a right quotient ring does it follow that R has a right quotient ring? Is it possible for R to be its own quotient ring and for there to be an infinite number of isomorphism types of simple right R-module? If R is its own quotient ring does it follow that $M_n(R)$ is its own quotient ring or even has a right quotient ring at all?

# 14 Simple Noetherian rings

Let R be a simple Noetherian ring of Krull dimension 1. We shall prove that every finitely-generated torsion R-module is cyclic, and that every large-enough finitely-generated torsion-free R-module has a free direct summand. Also if R has finite global dimension we shall prove that R is Morita-equivalent to an integral domain. These results are due to J.T. Stafford, and they can serve as an introduction to his work on stable structure of modules and related topics. We will give a brief explanation of the concept of Morita-equivalence which we hope will be enough for an understanding of its use here. The results marked with a "K", e.g. Theorem 14.6K, can be ignored by anyone who is not familiar with the general definition and properties of Krull dimension. The examples given in this chapter often have properties which are hard to establish, and we will state them without proof. In particular we shall give brief descriptions of the two Zalesskii-Neroslavskii examples, and, apart from the original sources, a convenient account of these in English can be found in a paper by K.R. Goodearl.

Notation. We will use dim(M) and Kdim(M) to denote the Goldie and Krull dimensions respectively of a module M. If we wish to indicate that M is being considered as a right (or left) R-module we will write $M_R$ (or ${}_RM$).

All we need to know about Krull dimension one is that if R is a simple left Noetherian ring then $\mathrm{Kdim}({}_RR) = 1$ if and only if R is not Artinian and R/K is Artinian for each essential left ideal K of R. When we say that an R-module is torsion (or torsion-free) we mean that it is torsion (or torsion-free) with respect to the regular elements of R.

__Theorem 14.1__. (Stafford) Let R be a simple left Noetherian ring with $\text{Kdim}({}_R R) = 1$, let f be a non-zero element of R, and let M be a torsion left R-module. Assume that $M = Ra + Rb$ for some $a, b \in M$, then $M = R(a + fxb)$ for some $x \in R$.

__Proof__. Let $m \in M$ then $cm = 0$ for some regular element c of R. Hence Rm is a homomorphic image of the Artinian module R/Rc so that Rm is Artinian. Therefore M is Artinian. The result is trivial if $b = 0$. Assume that $b \neq 0$. There exists $u \in R$ such that Rub is a simple submodule of Rb. Because the length of Rb/Rub is less than that of Rb, we may suppose by induction that $M = R(a + fyb) + Rub$ for some $y \in R$. There is a regular element d of R such that $d(a + fyb) = 0$. Because R is simple and $df \neq 0$ we have $RdfR = R$. Thus $Rub = RdfRub$ so that $dfzub \neq 0$ for some $z \in R$. We have $Rdfzub = Rub$.

Set $x = y + zu$. We have $d(a + fxb) = dfzub$ so that $Rdfzub \subseteq R(a + fxb)$. Thus $Rub \subseteq R(a + fxb)$. Also $a + fyb = a + fxb - fzub$ so that $R(a + fyb) \subseteq R(a + fxb)$. Therefore $M = R(a + fxb)$. □

__Corollary 14.2__. Let R be as in 14.1 and let M be a torsion left R-module. Assume that $M = Ra + Rb_1 + \ldots + Rb_n$ and let $f_1, \ldots, f_n$ be non-zero elements of R, then $M = R(a + f_1x_1b_1 + \ldots + f_nx_nb_n)$ for some $x_i \in R$.

__Corollary 14.3__. Let R be as in 14.1 then every finitely-generated torsion left R-module is cyclic.

__Corollary 14.4__. Let R be as in 14.1 then every left ideal of R can be generated by two elements.

__Proof__. Because every left ideal is a direct summand of an essential left ideal, it is enough to show that an arbitrary essential left ideal K of R can be generated by two elements. By Goldie's theorem we know that K contains a regular element c. Because R satisfies the left Ore condition, we also know that if $k \in K$ then $dk \in Rc$ for some regular element d of R. Thus K/Rc is a torsion module. Therefore K/Rc is cyclic (14.3), i.e. $K = Ra + Rc$ for some $a \in R$. □

Corollary 14.5. Let R be as in 14.1, let c be a regular element of R and let f be a non-zero element of R, then $R = Rc + Rfx$ for some $x \in R$.

Proof. Set $M = R/Rc$ and apply 14.1 with $a = 0 + Rc$ and $b = 1 + Rc$. □

Theorem 14.6K. (Stafford) Let R be a simple left Noetherian ring with $\mathrm{Kdim}({}_RR) = n$ where n is a positive integer. Let $f_1,\ldots,f_n$ be non-zero elements of R, and let M be a torsion left R-module of the form $M = Ra_1 + \ldots + Ra_n + Rb$, then there are elements $x_i$ of R such that $M = R(a_1 + f_1x_1b) + \ldots + R(a_n + f_nx_nb)$.

Proof. Assume that $b \neq 0$ and let $u \in R$ be such that Rub is a critical submodule of Rb. By induction on the length of a series of cyclic critical modules for Rb, there are elements $y_i$ of R such that $M = R(a_1 + f_1y_1b) + \ldots + R(a_n + f_ny_nb) + Rub$. There is a regular element d of R such that $d(a_i + f_iy_ib) = 0$ for all i. Because $Rdf_1R = R$ we have $df_1z_1ub \neq 0$ for some $z_1 \in R$. Thus $Rdf_1z_1ub$ is a non-zero submodule of the critical module Rub so that $\mathrm{Kdim}(Rub/Rdf_1z_1ub) < \mathrm{Kdim}(M)$. Also $Rub/Rdf_1z_1ub = Rk_2 + \ldots + Rk_n + Rb'$ where $k_i = 0$ for all i and $b' = ub + Rdf_1z_1ub$. Therefore by induction on Krull dimension we have $Rub = Rdf_2z_2ub + \ldots + Rdf_nz_nub + Rdf_1z_1ub$ for some $z_i \in R$. For each i set $x_i = y_i + z_iu$, then as in the proof of 14.1 we have $M = R(a_1 + f_1x_1b) + \ldots + R(a_n + f_nx_nb)$. □

Corollary 14.7K. Let R be as in 14.6K then every finitely-generated torsion left R-module can be generated by n elements and every left ideal of R can be generated by n + 1 elements.

Example 14.8. (The Weyl algebras) Let K be a field of characteristic 0 and let n be a positive integer, then the nth Weyl algebra over K, denoted $A_n(K)$ or $A_n$, is the K-algebra generated by elements $x_1,\ldots,x_n$, $y_1,\ldots,y_n$ subject to the relations $x_ix_j = x_jx_i$ and $y_iy_j = y_jy_i$ for all i and j, $x_iy_j = y_jx_i$ if $i \neq j$, and $x_iy_i - y_ix_i = 1$ for all i. It is known that $A_n$ is a simple left and right Noetherian integral domain of Krull dimension n and global dimension n. □

Example 14.9. (Zalesskii and Neroslavskii) Let K be the field of real numbers and let A be the first Weyl algebra over K. Thus A is the K-algebra with generators x and y such that xy - yx = 1, and A is a simple Noetherian domain. Let h be the K-automorphism of A given by $x^h = -x$ and $y^h = -y$. Note that $h^2 = 1$. Let R be the twisted group ring over A of the cyclic group generated by h; the elements of R can be expressed uniquely in the form a + hb with a, b $\in$ A with the usual addition and multiplication subject to the rules that $h^2 = 1$ and $ah = ha^h$ for all a $\in$ A. It can be shown that R is a simple left and right Noetherian hereditary ring of Krull dimension 1, but R is not a domain because (1 + h)(1 - h) = 0. Also R is not of the form $M_n(D)$ where D is an integral domain and n is a positive integer. For future reference we note that it is easy to show that dim(R) = 2. Set e =½(1 + h) then e is idempotent and eR is a uniform right ideal. Hence eRe is an integral domain which is Morita-equivalent to R. □

The next major aim is to show that if R is a simple Noetherian ring of Krull dimension 1 and M is a finitely-generated torsion-free right R-module with $\dim(M) \geq \dim(R) + 1$, then R is a direct summand of M.

Lemma 14.10. Let R be a semi-prime right Goldie ring with right quotient ring Q. Let K be a submodule of a right R-module M and let $f:K \to Q$ be a right R-module homomorphism, then f can be extended to a right R-module homomorphism $g:M \to Q$.

Proof. There is a submodule C of M such that $C \cap K = 0$ and $C \oplus K$ is essential in M. By setting $f(c) = 0$ for all $c \in C$ we can assume that f is defined on $C \oplus K$. Therefore without loss of generality we may assume that K is essential in M. Let $x \in M$ then $xc \in K$ for some regular element c or R (1.1 and Goldie's theorem). Set $g(x) = f(xc)c^{-1}$. To show that g(x) is well-defined assume also that $xd \in K$ where d is regular. By the right Ore condition we have $cs = dt$ for some $s, t \in R$ with s regular. Because $dt = cs$ it is clear that t is right regular and hence regular (1.13). Set $w = cs = dt$ then w is a unit of Q and $f(xc)c^{-1}w = f(xc)s = f(xcs) = f(xdt) = f(xd)t = f(xd)d^{-1}w$. Therefore $f(xd)d^{-1} = f(xc)c^{-1}$ so that g is well-defined. We leave it as an exercise to show that g is a right R-module homomorphism. □

Lemma 14.11. Let R be a semi-prime left and right Goldie ring and let U be a finitely-generated torsion-free uniform right R-module, then U is isomorphic to a right ideal of R.

Proof. Let Q be the quotient ring of R. Let u be a non-zero element of U and set $r(u) = \{r \in R: ur = 0\}$. We have $uR \cong R/r(u)$. Because U is torsion-free we know that $r(u)$ is not an essential right ideal of R. Hence $I \cap r(u) = 0$ for some non-zero right ideal I of R. Thus I is isomorphic to a submodule of $R/r(u)$. Hence there is a non-zero submodule V of U and an embedding $f:V \to R$. We can extend f to a homomorphism $g:U \to Q$ (14.10). Clearly g is also an embedding because V is essential in U. Thus there is a finitely-generated right R-submodule W of Q such that $W \cong U$. Suppose that $W = q_1R + \ldots + q_nR$. Because Q is the left quotient ring of R, there is a regular element c of R such that $cq_i \in R$ for all i. Therefore $cW \subseteq R$ and $cW \cong W$. Hence $U \cong cW$ and cW is a right ideal of R. □

<u>Lemma 14.12</u>. Let R be a prime left and right Goldie ring and let U and V be finitely-generated torsion-free uniform right R-modules, then U is isomorphic to a submodule of V.

<u>Proof</u>. Without loss of generality we may suppose that U and V are right ideals of R (14.11). We have $VU \neq 0$ so that $vU \neq 0$ for some $v \in R$. Define $f:U \to V$ by $f(u) = vu$ for all $u \in U$. Then f is a right R-module homomorphism. We have $\mathrm{Ker}(f) \neq U$. Because R is right non-singular (1.6) it follows easily that Ker(f) is not essential in U. Therefore $\mathrm{Ker}(f) = 0$. □

We recall that dim(M) denotes the Goldie dimension of a module M. Let R be a prime left and right Goldie ring then the quotient ring Q of R is of the form $M_k(D)$ for some positive integer k and division ring D. We have $\dim(R_R) = \dim(Q_Q) = k = \dim({}_QQ) = \dim({}_RR)$. Thus the left and right Goldie dimensions of R are equal and we just write dim(R). It is easy to extend this argument to semi-prime left and right Goldie rings.

<u>Theorem 14.13</u>. (Stafford) Let R be a simple left Noetherian right Goldie ring with $\mathrm{Kdim}({}_RR) = 1$ and let M be a finitely-generated torsion-free right R-module with $\dim(M) \geq \dim(R) + 1$, then R is isomorphic to a direct summand of M.

<u>Proof</u>. Set $k = \dim(R)$. There are finitely-generated uniform submodules $U_1, \ldots, U_{k+1}$ of M such that the sum $U_1 + \ldots + U_{k+1}$ is direct. Also there

are finitely-generated uniform right ideals $V_1,\ldots,V_k$ of R whose sum is direct. Set $I = V_1 \oplus \ldots \oplus V_k$ then I is an essential right ideal of R. Because $U_i$ is isomorphic to a submodule of $V_i$ for $1 \le i \le k$ (14.12) we may without loss of generality assume that $U_i \cong V_i$. Thus there is a homomorphism $g:(U_1 \oplus \ldots \oplus U_{k+1}) \to I$ such that $g(U_{k+1}) = 0$ and $g(U_i) = V_i$ for $1 \le i \le k$. Let Q be the quotient ring of R. By 14.10 we may assume that g has been extended to a right R-module homomorphism from M into Q. Similarly there is a homomorphism $h:M \to Q$ such that $h(U_1 \oplus \ldots \oplus U_k) = 0$ and $h(U_{k+1}) \neq 0$. Because g(M) and h(M) are finitely-generated right R-submodules of Q there is a regular element d of R such that $dg(M) \subseteq R$ and $dh(M) \subseteq R$ (cf. proof of 14.11). Replacing g by dg etc. we may suppose that $g(M) \subseteq R$ and $h(M) \subseteq R$.

There is a regular element c of R such that $c \in I$. We have $c = g(u)$ for some $u \in U_1 \oplus \ldots \oplus U_k$. Let $v \in U_{k+1}$ with $h(v) \neq 0$ and set $f = h(v)$. Note that $g(v) = 0 = h(u)$. We have $1 = yc + zfx$ for some $x, y, z \in R$ (14.5). Define $s:M \to R$ by $s(m) = yg(m) + zh(m)$ for all $m \in M$. We have $s(u + vx) = yg(u + vx) + zh(u + vx) = yg(u) + zh(v)x = yc + zfx = 1$ so that $s(M) = R$. Therefore R is isomorphic to a direct summand of M. □

Corollary 14.14. Let R be as in 14.13 and let M be a finitely-generated right R-module with $\rho(M) \ge \dim(R) + 1$, then R is isomorphic to a direct summand of M (here $\rho(M)$ denotes the rank of M as defined in Chapter 2).

Proof. Let T(M) be the torsion submodule of M then M/T(M) is torsion-free and $\dim(M/T(M)) = \rho(M) \ge \dim(R) + 1$. Therefore R is a homomorphic image of M/T(M) (14.13) and hence also of M. □

Theorem 14.15K. (Stafford) Let R be a simple left Noetherian right Goldie ring and let $\mathrm{Kdim}({}_RR) = n$ where n is a positive integer. Let M be a finitely-generated right R-module with $\rho(M) \ge \dim(R) + n$ then R is isomorphic to a direct summand of M.

<u>Proof</u>. Without loss of generality we may suppose that M is torsion-free (cf. proof of 14.14). We shall show how to modify the proof of 14.13. Set $k = \dim(R)$ and let $U_1,\ldots,U_{k+n}$ be finitely-generated uniform submodules of M whose sum is direct. Because $U_1 \oplus \ldots \oplus U_k$ is isomorphic to an essential right ideal of R, there is a homomorphism $g:M \to R$ such that $g(U_1 \oplus \ldots \oplus U_k)$ contains a regular element c of R and $g(U_{k+1}) = \ldots = g(U_{k+n}) = 0$. Also for $1 \le i \le n$ there is a homomorphism $h_i:M \to R$ such that $h_i(U_j) = 0$ if and only if $j \neq k + i$. Let $v_i \in U_{k+i}$ with $h_i(v_i) \neq 0$ and set $f_i = h_i(v_i)$. Let $u \in U_1 \oplus \ldots \oplus U_k$ with $g(u) = c$. We have $1 = yc + z_1f_1x_1 + \ldots + z_nf_nx_n$ for some $x_i$, $y$, $z_i \in R$ (14.6K applied to R/Rc). Define $s:M \to R$ by $s(m) = yg(m) + z_1h_1(m) + \ldots + z_nh_n(m)$ for all $m \in M$, then $s(u + v_1x_1 + \ldots + v_nx_n) = 1$. □

We next give a brief discussion of Morita-equivalence. We will give only the information which we need and will not attempt to make the discussion comprehensive. Two rings R and S being Morita-equivalent is a generalisation of the case where $S \cong M_n(R)$ for some positive integer n (i.e. S is the endomorphism ring of a free R-module of finite rank). We will often not distinguish between isomorphic objects.

<u>Notation 14.16</u>. Let R be a ring and let P be a finitely-generated projective right R-module. Let S be the endomorphism ring of $P_R$. If $s \in S$ and $p \in P$ we shall write sp rather than s(p). In this way P is a left S-module. Set $P^* = \mathrm{Hom}(P_R,R_R)$. If $f \in P^*$ and $p \in P$ we write fp rather than f(p), and by $P^*P$ we mean the subset of R consisting of all sums of finitely-many terms of the form fp. Clearly $P^*P$ is an ideal of R. Let $f \in P^*$ and $p \in P$, then by pf we mean the element of S defined by $pf.q = p.fq$ for all $q \in P$ (note that $fq \in R$). As above we can use this to give a meaning to $PP^*$, and $PP^*$ is an

ideal of S. By the dual basis lemma (8.20) there exist $p_1,\ldots,p_n \in P$ and $f_1,\ldots,f_n \in P^*$ such that $q = p_1f_1q + \ldots + p_nf_nq$ for all $q \in P$. Thus $p_1f_1 + \ldots + p_nf_n = 1$ so that $1 \in PP^*$. Therefore $PP^* = S$. Hence $P=SP = PP^*P$ so that $P^*P \neq 0$, and $(P^*P)^2 = P^*PP^*P = P^*SP = P^*P$. Thus $P^*P$ is a non-zero idempotent ideal of R (called the trace ideal of P), but it is not always true that $P^*P = R$. For example, if R is the ring of 2 by 2 upper triangular matrices over a field and $P = e_{22}R$ then $P^* = Re_{22} = P^*P$. However if R is simple then clearly $P^*P = R$.

With the notation of 14.16 and with the extra assumption that $P^*P = R$ we say that R and S are *Morita-equivalent*. From now on we will omit detailed verifications. In a natural way $P^*$ can be made into a left R-module and a right S-module, and the relationship between R, $P^*$ and S is the same as that between S, P and R. Thus Morita-equivalence is a symmetric relation. We have $P \oplus Q = R^n$ for some projective right R-module Q and positive integer n. The endomorphism ring of $P \oplus Q$ is $M_n(R)$. Let e be the projection of $P \oplus Q$ onto P along Q then $S = eM_n(R)e$ and $M_n(R)eM_n(R) = M_n(R)$. For convenience we shall write $R_n$ rather than $M_n(R)$, etc. In fact R and S are Morita-equivalent if and only if there exist a positive integer n and an idempotent e such that $S = eR_ne$ and $R_n = R_neR_n$. This characterisation gives one way of proving that Morita-equivalence is transitive. For suppose that $S = eR_ne$ with $R_n = R_neR_n$ and that $T = fS_kf$ with $S_k = S_kfS_k$ where e and f are idempotent. Let e' be the k by k matrix with e in each diagonal position and 0's elsewhere then $S_k = (eR_ne)_k = e'R_{kn}e'$. But $f \in S_k$ so that $f = e'f = fe'$. Hence $T = fR_{kn}f$ and $R_{kn} = (R_n)_k = (R_neR_n)_k = R_{kn}e'R_{kn} = R_{kn}e'R_{kn}e'R_{kn} = R_{kn}S_kR_{kn} = R_{kn}S_kfS_kR_{kn} = R_{kn}fR_{kn}$. Therefore T and R are Morita-equivalent. Although we are avoiding category theory, we should at least mention that rings R and S are Morita-equivalent if and only if the category of right

R-modules is equivalent to the category of right S-modules.

A property X is said to be *Morita-invariant* if, whenever R and S are Morita-equivalent rings, then R satisfies X if and only if S satisfies X. For example, with the notation of 14.16 suppose that R is simple and let I be a non-zero ideal of S, then P*IP is a non-zero ideal of R. Hence P*IP = R so that I = SIS = PP*IPP* = PRP* = PP* = S. Therefore S is also simple. Similarly, let A and B be right ideals of S such that AP = BP then A = AS = APP* = BPP* = BS = B. From this it follows easily for example that if S is Morita-equivalent to R and R is right Noetherian then so also is S.

The following are some examples of properties which are Morita-invariant: being simple, prime, semi-prime, right or left Artinian or Noetherian, having Krull dimension 1, finite Goldie dimension, global dimension n. The following are not Morita-invariant: being an integral domain or a p.p. ring or a principal right ideal ring, having right Goldie dimension n where n is a fixed positive integer.

We now turn to the final topic of this chapter. It is well-known that a simple Artinian ring is a matrix ring over a division ring. Let R be a simple Noetherian ring, then it is natural to ask:

(1) Is R always a matrix ring over an integral domain?

(2) Is R always Morita-equivalent to an integral domain?

(3) If R is not an integral domain does R always have a non-trivial idempotent element?

Clearly if the answer to (2) or (3) is "No" then the answer to (1) is also "No". The Zalesskii-Neroslavskii example given in 14.9 shows that the answer to (1) is "No", but in that case R is Morita-equivalent to a domain and R has a non-trivial idempotent element (e.g. $\frac{1}{2}(1 + h)$). However the following example (whose properties are even harder to verify) shows that the answer to

(3) is "No", and J.T. Stafford proved (14.18) that this example also shows that the answer to (2) is "No".

Example 14.17. (Zalesskii-Neroslavskii) We shall construct a simple Noetherian ring $R$ which is not a domain and which has no idempotent elements except 0 and 1. Let $K$ be a field of characteristic 2 and let $x$ and $y$ be indeterminates. Let $L$ be the field of rational functions in $y$ over $K$. The polynomial ring $L[x]$ is hereditary and hence so also is the partial quotient ring $R_1$ of $L[x]$ with respect to the powers of $x$ (alternatively $R_1$ can be thought of as the group algebra over $L$ of the infinite cyclic group generated by $x$). Let $g$ be the $L$-automorphism of $R_1$ given by $x^g = yx$. Let $R_2$ be the twisted group ring over $R_1$ of the infinite cyclic group generated by $g$. Because no proper ideal of $R_1$ is invariant under $g$, it follows by a result of A. Shamsuddin that $R_2$ also is hereditary and hence has Krull dimension 1. Let $h$ be the $L$-automorphism of $R_2$ given by $x^h = x^{-1}$ and $g^h = g^{-1}$. Note that $h^2 = 1$. Let $R$ be the twisted group ring over $R_2$ of the cyclic group generated by $h$. Thus the elements of $R$ can be expressed uniquely in the form $a + hb$ with $a, b \in R_2$, and $ah = ha^h$ for all $a \in R_2$. Clearly $R$ is not a domain because $(1 + h)(1-h)=0$. It can be shown that $R$ is a simple left and right Noetherian ring with no proper idempotents. It is necessary for $K$ to have characteristic 2 for otherwise $\frac{1}{2}(1 + h)$ would be an idempotent element of $R$. Because $R$ is finitely-generated as an $R_2$-module and $R_2$ has Krull dimension 1 so also has $R$. Also $r(1 + h) = (1 + h)R$ so that there is an exact sequence $0 \to (1+h)R \to R \to (1+h)R \to 0$ and it follows that $R$ has infinite global dimension (cf. 14.19). □

Theorem 14.18. (Stafford) Let $R$ be the ring given in 14.17 then $R$ is not Morita-equivalent to an integral domain.

_Proof_. Suppose that there is a finitely-generated projective right R-module P such that the endomorphism ring S of P is a domain. Because S is a Noetherian domain we have dim(S) = 1. We know that R has no non-trivial idempotents so that $_RR$ is indecomposable. Set P* = Hom(P,R) as in 14.16. Let A and B be submodules of $_SP$ such that P = A ⊕ B then R = P*A ⊕ P*B. Hence either P*A = 0 or P*B = 0, i.e. either A = 0 or B = 0. Therefore $_SP$ is indecomposable. But dim(S) + Kdim($S_S$) = 2. Hence dim($_SP$) < 2 (14.13), i.e. dim($_SP$) = 1. Therefore dim(R) = 1. Thus every non-zero left (right) ideal of R is essential and so has zero right (left) annihilator (1.6). From this it follows easily that R is an integral domain, which is a contradiction. □

_Theorem 14.19_. (Stafford) Let R be a simple left and right Noetherian ring with Kdim($_RR$) = 1. Assume also that R has finite global dimension, then R is Morita-equivalent to an integral domain.

_Proof_. Amongst all rings which are Morita-equivalent to R choose S such that dim(S) is as small as possible. We wish to show that S is a domain, i.e. that dim(S) = 1 (cf. end of proof of 14.18).

Let P be a finitely-generated indecomposable projective right S-module and set k = dim(S). Because P is indecomposable we have dim(P) ≤ k (14.13). Let T be the endomorphism ring of P then T is Morita-equivalent to S and hence also to R. Also we have dim(T) = dim($P_S$). Hence by the minimality of dim(S) we have dim(P) ≥ k. Therefore dim(P) = dim(S) for each finitely-generated indecomposable projective right S-module P. It is easy to modify the technique of R. Walker given in the proof of 10.4 to show that dim(S) divides $\rho(M)$ for each finitely-generated right S-module M. Therefore dim(S) = 1. □

_Theorem 14.20K_. (Stafford) Let R be a simple left and right Noetherian ring with Kdim($_RR$) = n where n is a positive integer. Suppose also that R has finite global dimension. Then R is Morita-equivalent to a ring S with $\dim(S) \leq n$.

Proof. As in the proof of 14.19 we choose S Morita-equivalent to R with $\dim(S)$ as small as possible. Set $k = \dim(S)$ and assume that $n < k$. Let P be any finitely-generated projective right S-module (we do not consider the zero module to be projective). As in 14.19 the minimality of $\dim(S)$ gives $\dim(P) \geq k$. If $\dim(P) \geq k + n$ then $P \cong S \oplus P'$ for some projective right S-module P' (14.15K). Because $\dim(P') \geq k$ we have $\dim(P) \geq 2k$. Thus it is not possible to have $k + n \leq \dim(P) < 2k$. If P is any finitely-generated indecomposable projective right S-module then $k \leq \dim(P) < k + n$.

Set $H = \{a \in Z: 0 \leq a < n$ and $k + a = \dim(P)$ for some finitely-generated projective right S-module $P\}$. For the rest of this proof P, Q, X, Y, etc. will denote finitely-generated projective right S-modules. If P is indecomposable then $\dim(P) = k + h$ for some $h \in H$. If H has no non-zero elements then we can prove as in 14.19 that S is a domain.

Now let a be the smallest positive element of H, let $h \in H$, and fix P and Q with $\dim(P) = k + a$ and $\dim(Q) = k + h$. We have $\dim(P \oplus Q) = 2k + a+h \geq k+n$. Hence $P \oplus Q = S \oplus X$ with $\dim(X) = k + a + h$ (14.15K). Thus if $a + h < n$ then $a + h \in H$. Suppose that $a + h \geq n$ then $X = S \oplus Y$ with $\dim(Y) = a + h$. Hence $a + h \geq k$. Thus $\dim(Y) = k + (a + h - k)$ and $a + h - k \geq 0$. We cannot have $a + h - k \geq a$, so that $a + h - k = 0$.

Thus for all $h \in H$ we have either $a + h \in H$ or $a + h = k$. Let t be the largest positive integer such that $ta \in H$ then $a + ta = k$. Hence a divides k. Also $ta < n < k = (1 + t)a$ so that $a \neq 1$. Now let $h \in H$. Let u be the largest non-negative integer such that $ua + h \in H$ then $(1 + u)a + h = k$ so that a divides h and hence also $k + h$.

We know now that a divides $\dim(P)$ for each finitely-generated indecomposable projective right S-module P. Because S has finite global dimension it follows that a divides $\rho(M)$ for each finitely-generated right S-module M. Therefore a = 1, which is a contradiction. □

Remarks

(1) For further details and generalisations see the relevant papers by J.T. Stafford.

(2) No example is known of a simple left and right Noetherian ring with a right ideal which cannot be generated by two elements.

(3) Every known example of a simple left and right Noetherian ring of Krull dimension 1 which has finite global dimension in fact has global dimension 1.

(4) It is not known whether a simple left and right Noetherian ring of finite global dimension is always Morita-equivalent to a domain.

(5) It is not known whether a simple left and right Noetherian ring of left Krull dimension 1 always has right Krull dimension 1.

# References

1. S.A. Amitsur, "Prime rings having polynomial identities with arbitrary coefficients", Proc. London Math. Soc., (3) 17 (1967), 470-486.

2. G.M. Bergman, "Some examples in PI ring theory", Israel J. Math., 18 (1974), 257-277.

3. G.M. Bergman, "A prime ring of right and left Goldie dimension 1 having an infinite chain of right annihilator ideals", unpublished note.

4. S.M. Bhatwadekar, "On the global dimension of some filtered algebras", J. London Math. Soc., (2) 13 (1976), 239-248.

5. M. Boratyński, "A change of rings theorem and the Artin-Rees property", Proc. Amer. Math. Soc., 53 (1975), 307-310.

6. K.A. Brown, "The singular ideals of group rings", Quart. J. Math. Oxford, (2) 28 (1977), 41-60.

7. K.A. Brown, T.H. Lenagan and J.T. Stafford, "Weak ideal invariance and localisation", to appear in J. London Math. Soc.

8. K.A. Brown, T.H. Lenagan and J.T. Stafford, "Stable structure and K-theory of some Noetherian group rings", to appear in Proc. London Math. Soc.

9. K.A. Brown, C.R. Hajarnavis and A.B. MacEacharn, "Noetherian rings of finite global dimension", to appear in Proc. London Math. Soc.

10. K.A. Brown, "Module extensions over Notherian rings".

11. G. Cauchon, "Les T-anneaux et la condition de Gabriel", C.R. Acad. Sci. Paris, t.277 (1973), Séries A, 1153-1156.

12. G. Cauchon, "Sur l'intersection des puissances du radical d'un T-anneau Noethérien", C.R. Acad. Sci. Paris, t.279 (1974), Séries A, 91-93.

13. G. Cauchon, "Les T-anneaux, la condition (H) de Gabriel et ses conséquences", Comm. in Algebra, 4 (1976), 11-50.

14. G. Cauchon, "Anneaux semi-premiers, Noethériens, à identités polynomiales", Bull. Soc. Math. France, 104 (1976), 99-111.

15. S.U. Chase, "A generalization of the ring of triangular matrices", Nagoya Math. J., 18 (1961), 13-25.

16. A.W. Chatters and C.R. Hajarnavis, "Non-commutative rings with primary decomposition", Quart. J. Math. Oxford, (2) 22 (1971), 73-83.

17. A.W. Chatters, "The restricted minimum condition in Noetherian hereditary rings", J. London Math. Soc., (2) 4 (1971), 83-87.

18. A.W. Chatters and S.M. Ginn, "Localisation in hereditary rings", J. of Algebra, 22 (1972), 82-88.

19. A.W. Chatters, "A decomposition theorem for Noetherian hereditary rings". Bull. London Math. Soc., 4 (1972), 125-126.

20. A.W. Chatters, "A non-singular Noetherian ring need not have a classical quotient ring", J. London Math. Soc., (2) 10 (1975), 66-68.

21. A.W. Chatters, "Two results on p.p. rings", Comm. in Algebra, 4 (1976), 881-891.

22. A.W. Chatters and C.R. Hajarnavis, "Rings in which every complement right ideal is a direct summand", Quart. J. Math. Oxford, (2) 28 (1977), 61-80.

23. A.W. Chatters, C.R. Hajarnavis and N.C. Norton, "The Artin radical of a Noetherian ring", J. Australian Math. Soc., 23 (1977), 379-384.

24. A.W. Chatters and P.F. Smith, "A note on hereditary rings", J. of Algebra, 44 (1977), 181-190.

25. A.W. Chatters, "A note on Noetherian orders in Artinian rings", Glasgow Math. J., 20 (1979), 125-128.

26. A.W. Chatters, "Three examples concerning the Ore condition in Noetherian rings", to appear in Proc. Edinburgh Math. Soc.

27. A.W. Chatters and S.M. Khuri, "Endomorphism rings of modules over non-singular CS rings", to appear in J. London Math. Soc.

28. A.W. Chatters, A.W. Goldie, C.R. Hajarnavis and T.H. Lenagan, "Reduced rank in Noetherian rings", to appear in J. of Algebra.

29. A.W. Chatters and J.C. Robson, "Decomposition of orders in semi-primary rings", to appear in Comm. in Algebra.

30. P.M. Cohn, "On a class of binomial extensions", Illinois J. Math., 10 (1966), 418-424.

31. P.M. Cohn, "Skew field constructions", London Math. Soc. Lecture Notes, 27 (1977), Cambridge University Press.

32. R.R. Colby and E.A. Rutter Jr., "Generalisations of QF3-algebras", Trans. Amer. Math. Soc., 153 (1971), 371-386.

33. J.H. Cozzens, "Homological properties of the ring of differential polynomials", Bull. Amer. Math. Soc., 76 (1970), 75-79.

34. M.G. Deshpande, "Structure of right subdirectly irreducible rings I", J. of Algebra, 17 (1971), 317-325.

35. J. Dixmier, "Algèbres enveloppantes", Cahiers Scientifiques, Fascicule XXXVII, Gauthier-Villars, (1974).

36. M. Djabali, "Anneau de fractions d'un J-anneau", Canadian J. Math., 20 (1968), 182-202.

37. M. Djabali, "Anneaux de fractions "presque classiques" Artiniens", J. of Algebra, 16 (1972), 116-128.

38. D. Eisenbud and J.C. Robson, "Modules over Dedekind prime rings", J. of Algebra, 16 (1970), 67-85.

39. D. Eisenbud and J.C. Robson, "Hereditary Noetherian prime rings", J. of Algebra, 16 (1970), 86-104.

40. D. Eisenbud and P. Griffith, "Serial rings", J. of Algebra, 17 (1971), 389-400.

41. C. Faith, "Rings with ascending condition on annihilators", Nagoya Math. J., 27 (1966), 179-191.

42. K.L. Fields, "On the global dimension of residue rings", Pacific J. Math., 32 (1970), 345-349.

43. J.W. Fisher, "On the nilpotency of nil subrings", Canadian J. Math., 22 (1970), 1211-1216.

44. J.W. Fisher, "Nil subrings with bounded indices of nilpotency", J. of Algebra, 19 (1971), 509-516.

45. J.W. Fisher, "Nil subrings of endomorphism rings of modules", Proc. Amer. Math. Soc., 34 (1972), 75-78.

46. J.W. Fisher, "Structure of semi-prime PI-rings I", Proc. Amer. Math. Soc., 39 (1973), 465-467.

47. J.W. Fisher, "Finiteness conditions for projective and injective modules", Proc. Amer. Math. Soc., 40 (1973), 389-394.

48. E. Formanek and A.V. Jategaonkar, "Subrings of Noetherian rings", Proc. Amer. Math. Soc., 46 (1974), 181-186.

49. J. Fuelberth and J. Kuzmanovich, "The structure of semi-primary and Noetherian hereditary rings", Trans. Amer. Math. Soc., 212 (1975), 83-111.

50. J. Fuelberth, E. Kirkman and J. Kuzmanovich, "Hereditary module-finite algebras", J. London Math. Soc., (2) 19 (1979), 268-276.

51. S.M. Ginn and P.B. Moss, "Finitely embedded modules over Noetherian rings", Bull. Amer. Math. Soc., 81 (1975), 709-710.

52. S.M. Ginn, "A counter-example to a theorem of Kurshan", J. of Algebra, 40 (1976), 105-106.

53. S.M. Ginn and P.B. Moss, "A decomposition theorem for Noetherian orders in Artinian rings", Bull. London Math. Soc., 9 (1977), 177-181.

54. A.W. Goldie, "The structure of prime rings under ascending chain conditions", Proc. London Math. Soc., (3) 8 (1958), 589-608.

55. A.W. Goldie, "Semi-prime rings with maximum condition". Proc. London Math. Soc., (3) 10 (1960), 201-220.

56. A.W. Goldie, "Torsion-free modules and rings", J. of Algebra, 1 (1964), 268-287.

57. A.W. Goldie, "Localisation in non-commutative Noetherian rings", J. of Algebra, 5 (1967), 89-105.

58. A.W. Goldie and L.W. Small, "A note on rings of endomorphisms", J. of Algebra, 24 (1973), 392-395.

59. K.R. Goodearl, "Global dimension of differential operator rings", Proc. Amer. Math. Soc., 45 (1974), 315-322.

60. K.R. Goodearl, "Global dimension of differential operator rings II", Trans. Amer. Math. Soc., 209 (1975), 65-85.

61. K.R. Goodearl, "Global dimension of differential operator rings III", J. London Math. Soc., (2) 17 (1978), 397-409.

62. K.R. Goodearl, "Simple Noetherian rings - the Zalesskii-Neroslavskii examples" in "Ring theory, Waterloo 1978", edited by David Handelman and John Lawrence, Lecture Notes in Mathematics No. 734 (1979), Springer-Verlag.

63. K.R. Goodearl, "Incompressible critical modules", to appear in Comm. in Algebra.

64. R. Gordon, "Semi-prime right Goldie rings which are direct sums of uniform right ideals", Bull. London Math. Soc., 3 (1971), 277-282.

65. R. Gordon and L.W. Small, "Piecewise domains", J. of Algebra, 23 (1972), 553-564.

66. R. Gordon, "Classical quotient rings of "PWD's", Proc. Amer. Math. Soc., 36 (1972), 39-46.

67. R. Gordon and J.C. Robson, "Krull dimension", Memoirs of the Amer. Math. Soc., 133 (1973).

68. R. Gordon, "Primary decomposition in right Noetherian rings", Comm. in Algebra, 2 (1974), 491-524.

69. R. Gordon, "Artinian quotient rings of FBN rings", J. of Algebra, 35 (1975), 304-307.

70. P. Griffith, "On the decomposition of modules and generalised left uniserial rings", Math. Ann., 184 (1970), 300-308.

71. C.R. Hajarnavis, "Orders in QF and QF2 rings", J. of Algebra, 19 (1971), 329-343.

72. C.R. Hajarnavis, "On Small's theorem", J. London Math. Soc., (2) 5 (1972), 596-600.

73. C.R. Hajarnavis and T.H. Lenagan, "Localisation in Asano orders", J. of Algebra, 21 (1972), 441-449.

74. C.R. Hajarnavis, "Non-commutative rings whose homomorphic images are self-injective", Bull. London Math. Soc., 5 (1973), 70-74.

75. C.R. Hajarnavis and N.C. Norton, "The one and half generator property in Noetherian rings", to appear in Comm. in Algebra.

76. R. Hart, "Simple rings with uniform right ideals", J. London Math. Soc., 42 (1967), 614-617.

77. I.N. Herstein and L.W. Small, "Nil rings satisfying certain chain conditions", Canadian J. Math., 16 (1964), 771-776.

78. I.N. Herstein, "A counter-example in Noetherian rings", Proc. Nat. Acad. Sci. (U.S.A.), 54 (1965), 1036-1037.

79. I.N. Herstein and L.W. Small, "Addendum to "Nil rings satisfying certain chain conditions" ", Canadian J. Math., 18 (1966), 300-302.

80. I.N. Herstein and L.W. Small, "Regular elements in P.I.-rings", Pacific J. Math., 46 (1971), 327-330.

81. A.V. Jategaonkar, "Left principal ideal domains", J. of Algebra, 8 (1968), 148-155.

82. A.V. Jategaonkar, "A counter-example in ring theory and homological algebra", J. of Algebra, 12 (1969), 418-440.

83. A.V. Jategaonkar, "Left principal ideal domains", Lecture Notes in Mathematics No. 123 (1970), Springer-Verlag.

84. A.V. Jategaonkar, "Jacobson's conjecture and modules over fully bounded Noetherian rings", J. of Algebra, 30 (1974), 103-121.

85. A.V. Jategaonkar, "Relative Krull dimension and prime ideals in right Noetherian rings", Comm. in Algebra, 2 (1974), 429-468.

86. A.V. Jategaonkar, "Principal ideal theorem for Noetherian P.I. rings", J. of Algebra, 35 (1975), 17-22.

87. A.V. Jategaonkar, "Certain injectives are Artinian", Proc. Kent State Conf., Lecture Notes in Mathematics No. 545 (1975), Springer-Verlag.

88. R.E. Johnson and L.S. Levy, "Regular elements in semi-prime rings", Proc. Amer. Math. Soc., 19 (1968), 961-963.

89. S. Jondrup, "Rings of quotients of some semiprime P.I. rings", Comm. in Algebra, 7 (1979), 279-286.

90. D.A. Jordan, "A left Noetherian, right Ore domain which is not right Noetherian", to appear in Bull. London Math. Soc.

91. J.W. Kerr, "Some examples of Goldie rings", Ph.D. thesis, University of California (San Diego), (1979).

92. G. Krause, T.H. Lenagan and J.T. Stafford, "Ideal invariance and Artinian quotient rings", J. of Algebra, 55 (1978), 145-154.

93. H. Kupisch, "Beiträge zur Theorie nichthalbeinfacher Ringe mit Minimalbedingung", J. Reine Angew. Math., 201 (1959), 100-112.

94. C. Lanski, "Nil subrings of Goldie rings are nilpotent", Canadian J. Math, 21 (1969), 904-907.

95. T.H. Lenagan, "Bounded Asano orders are hereditary", Bull. London Math. Soc., 3 (1971), 67-69.

96. T.H. Lenagan, "Bounded hereditary Noetherian prime rings", J. London Math. Soc., (2) 6 (1973), 241-246.

97. T.H. Lenagan, "Artinian ideals in Noetherian rings", Proc. Amer. Math. Soc., 51 (1975), 499-500.

98. T.H. Lenagan, "Artinian quotient rings of Macaulay rings", Proc. Kent State Conf., Lecture Notes in Mathematics No. 545 (1975), Springer-Verlag.

99. T.H. Lenagan, "Noetherian rings with Krull dimension one", J. London Math. Soc., (2) 15 (1977), 41-47.

100. T.H. Lenagan, "Reduced rank in rings with Krull dimension", to appear in the proceedings of the Antwerp ring theory conference 1978.

101. T.H. Lenagan and P.B. Moss, "K-symmetric rings", to appear in J. London Math. Soc.

102. L.S. Levy, "Torsion-free and divisible modules over non-integral domains", Canadian J. Math., 15 (1963), 132-151.

103. A.T. Ludgate, "A note on non-commutative Noetherian rings", J. London Math. Soc., (2) 5 (1972), 406-408.

104. J.C. McConnell, "The intersection theorem for a class of non-commutative Noetherian rings", Proc. London Math. Soc., (3) 17 (1967), 487-498.

105. J.C. McConnell, "Localisation in enveloping rings", J. London Math. Soc., 43 (1968), 421-428.

106. A.C. Mewborn and C.N. Winton, "Orders in self-injective semi-perfect rings", J. of Algebra, 13 (1969), 5-9.

107. G.O. Michier, "Asano orders", Proc. London Math. Soc., (3) 19 (1969), 421-443.

108. I.M. Musson, "Injective modules for group rings of polycyclic groups I".

109. I.M. Musson, "Injective modules for group rings of polycyclic groups II".

110. I.M. Musson, "Serial modules over enveloping algebras", unpublished note.

111. Y. Nouazé and P. Gabriel, "Idéaux premiers de l'algèbre enveloppante d'une algèbre de Lie nilpotente", J. of Algebra, 6 (1967), 77-99.

112. B.L. Osofsky, "Rings all of whose finitely-generated modules are injective", Pacific J. Math., 14 (1964), 645-650.

113. D.S. Passman, "The algebraic structure of group rings", Interscience (1977).

114. E.C. Posner, "Prime rings satisfying a polynomial identity", Proc. Amer. Math. Soc., 11 (1960), 180-183.

115. M. Ramras, "Orders with finite global dimension", Pacific J. Math., 50 (1974), 583-587.

116. J.C. Robson, "Artinian quotient rings", Proc. London Math. Soc., (3) 17 (1967), 600-616.

117. J.C. Robson, "Non-commutative Dedekind rings", J. of Algebra, 9 (1968), 249-265.

118. J.C. Robson, "Idealizers and hereditary Notherian prime rings", J. of Algebra, 22 (1972), 45-81.

119. J.C. Robson, "Decomposition of Noetherian rings", Comm. in Algebra, 1 (1974), 345-349.

120. J.E. Roseblade and P.F. Smith, "A note on the Artin-Rees property of certain polycyclic group algebras", Bull. London Math. Soc., 11 (1979), 184-185.

121. A. Rosenberg and D. Zelinsky, "Finiteness of the injective hull", Math. Z., 70 (1959), 372-380.

122. F.L. Sandomierski, "Non-singular rings", Proc. Amer. Math. Soc., 19 (1968), 225-230.

123. A. Shamsuddin, "A note on a class of simple Noetherian domains", J. London Math. Soc., (2) 15 (1977), 213-216.

124. J.C. Shepherdson, "Inverses and zero divisors in matrix rings", Proc. London Math. Soc., (3) 1 (1951), 71-85.

125. L.W. Small, "An example in Noetherian rings", Proc. Nat. Acad. Sci. (U.S.A.), 54 (1965), 1035-1036.

126. L.W. Small, "Hereditary rings", Proc. Nat. Acad. Sci. (U.S.A.), 55 (1966), 25-27.

127. L.W. Small, "Orders in Artinian rings", J. of Algebra, 4 (1966), 13-41.

128. L.W. Small, "Semi-hereditary rings", Bull. Amer. Math. Soc., 73 (1967), 656-658.

129. L.W. Small, "Orders in Artinian rings II", J. of Algebra, 9 (1968), 266-273.

130. L.W. Small and J.T. Stafford, in preparation.

131. P.F. Smith, "On the intersection theorem", Proc. London Math. Soc., (3) 21 (1970), 385-398.

132. P.F. Smith, "Localisation and the AR property", Proc. London Math. Soc., (3) 22 (1971), 39-68.

133. P.F. Smith, "Localisation in group rings", Proc. London Math. Soc., (3) 22 (1971), 69-90.

134. P.F. Smith, "Quotient rings of group rings", J. London Math. Soc., (2) 3 (1971), 645-660.

135. P.F. Smith, "On non-commutative regular local rings", Glasgow Math. J., 17 (1976), 98-102.

136. P.F. Smith, "Some rings which are characterised by their finitely generated modules", Quart. J. Math. Oxford, (2) 29 (1978), 101-109.

137. P.F. Smith, "Finitely embedded modules over group rings", Proc. Edinburgh Math. Soc., 21 (1978), 55-64.

138. P.F. Smith, "Rings characterised by their cyclic modules", Canadian J. Math., 31 (1979), 93-111.

139. P.F. Smith, "On the structure of certain P.P.-rings", Math. Z., 166 (1979), 147-157.

140. P.F. Smith, "The AR property and chain conditions in group rings", Israel J. Math., 32 (1979), 131-144.

141. P.F. Smith, "Certain rings are right hereditary".

142. J.T. Stafford, "Completely faithful modules and ideals of simple Noetherian rings", Bull. London Math. Soc., 8 (1976), 168-173.

143. J.T. Stafford, "Stable structure of noncommutative Noetherian rings", J. of Algebra, 47 (1977), 244-267.

144. J.T. Stafford, "Weyl algebras are stably free", J. of Algebra, 48 (1977), 297-304.

145. J.T. Stafford, "Stable structure of noncommutative Noetherian rings II", J. of Algebra, 52 (1978), 218-235.

146. J.T. Stafford, "A simple Noetherian ring not Morita equivalent to a domain", Proc. Amer. Math. Soc., 68 (1978), 159-160.

147. J.T. Stafford, "Module structure of Weyl algebras", J. London Math. Soc. (2) 18 (1978), 429-443.

148. J.T. Stafford, "Cancellation for nonprojective modules", in "Module theory", Lecture Notes in Mathematics No. 700 (1979), Springer-Verlag.

149. J.T. Stafford, "Morita equivalence of simple Noetherian rings", Proc. Amer. Math. Soc., 74 (1979), 212-214.

150. J.T. Stafford, "On the regular elements of Noetherian rings", to appear in the proceedings of the Antwerp ring theory conference 1978.

151. J.T. Stafford, "On a conjecture of Ramras", unpublished note.

152. J.T. Stafford, "Noetherian full quotient rings".

153. H. Tachikawa, "Quasi-Frobenius rings and generalisations", Lecture Notes in Mathematics No. 351 (1973), Springer-Verlag.

154. R. Walker, "Local rings and normalizing sets of elements", Proc. London Math. Soc., (3) 24 (1972), 27-45.

155. R.B. Warfield, "Serial rings and finitely presented modules", J. of Algebra, 37 (1975), 187-222.

156. R.B. Warfield, "Bezout rings and serial rings", Comm. in Algebra, 7 (1979), 533-545.

157. D.B. Webber, "Ideals and modules of simple Noetherian hereditary rings", J. of Algebra, 16 (1970), 239-242.

158. A.E. Zalesskii and O.M. Neroslavskii, "On simple Noetherian rings (Russian)", Isv. Acad. Nauk. BSSR, 5 (1975), 38-42.

159. A.E. Zalesskii and O.M. Neroslavskii, "There exists a simple Noetherian ring with divisors of zero but without idempotents (Russian)", Comm. in Algebra, 5 (1977), 231-234.

# Index